# ELECTRONIC NAVIGATION

## Colin Jones

Helmsman Books

First published in 1992 by
Helmsman Books, an imprint of
The Crowood Press Ltd
Ramsbury, Marlborough
Wiltshire SN8 2HR

**British Library Cataloguing in Publication Data**

A catalogue record for this book is available from the British Library

ISBN 1 85223 688 4

Line-drawings by Claire Upsdale-Jones
Photographs by Colin Jones

Typeset by Avonset, Midsomer Norton, Avon
Printed in Great Britain by Redwood Press Ltd, Melksham, Wilts

# CONTENTS

# PREFACE

This is a book for beginners and a 'post-graduate' seminar for those who have just done a bit of boating and who would now like to go further afield. My hope is that it will also be a refresher course for those whose navigational skills are suffering from the rustiness of disuse, but who would now like to catch up with modern electronics. Perhaps a coast-hopping reader, noting the safety and navigational assistance available to the modern skipper, might also be persuaded to move out of sight of the land and enjoy the adventure of arriving in another country under his own steam.

This is a sensation worth working for and I wish all such skippers as much pleasure in reading this book as I have derived from reliving past voyages and recalling old friends whilst writing it. The boats mentioned in the examples and anecdotes are a 6.8m (22.3ft) rigid hull inflatable, powered by a 90hp outboard motor, and an 8.75m (28.7ft) Colvic Watson motor sailer ketch. Both have taken me along many miles of coast and both have crossed the Channel to go even further south on a number of occasions.

Throughout the text, I have generally used the pronouns 'he' and 'him' to refer to both men and women. This writer's device has no sexist connotations, which would in any case be impossible on my own boat, where the co-skipper is a woman who shares the duties of boat handling and navigating completely. 'He' and 'him', therefore, are a writer's tool of convenience to avoid breaking up the text with repetitive 'he or she' or even 'he/she'. In this I have followed international and linguistic convention to improve the quality of the work and to maintain the ease of reading.

I wish you all the super sailing I have enjoyed with my partner and which we have both shared with solo navigators of both sexes and all those other marvellous boating and cruising people whom we have met *en route* from here to there.

*Colin Jones*

# INTRODUCTION

Electronic navigation is without doubt a major miracle, even when set against the other achievements of a century which has seen the development of everything from television to men on the moon. Pilotage technology at sea is miraculous because it is absolutely comprehensive in both time and space. No matter where you are in the world and irrespective of the time of day, or of the season of the year and with total disregard to weather conditions, there will be an electronic system available to tell you where you are and where you are going. In addition to all this science, electronic navigation is a lot of fun.

A major drawback of traditional navigation is that it is a cocktail of art and science. It also relies heavily on esti-mation and guesswork and is totally dependent on vision and visibility. Alas, there are periods of time when you cannot see 'a star to steer her by' for a week and even the sun will not come out to play with your sextant. However, when it does, a positional fix accurate to within a mile is an excellent result.

When a small ship is closing a coast without using some of the phenomenally accurate position-fixing devices recently available, adjustments to the course are usually made according to radio direction-finding information, and then refined as soon as a lighthouse, or other landmark is observed. The final approach to the des-tination relies on marker buoys and leading marks, backed up by a verifying glance at the tide level indicator. In theory, however, the electronically equipped navigator can do all these things in total darkness or in thick fog and be accurate to a few metres. Even allowing for electronic aberrations and human foibles, the practice now comes very close to the theory. Out at sea and along the coast, total electronic pilotage works well and it is even improving rapidly for zero visibility manoeuvring in crowded harbours.

For example, it is already possible to set

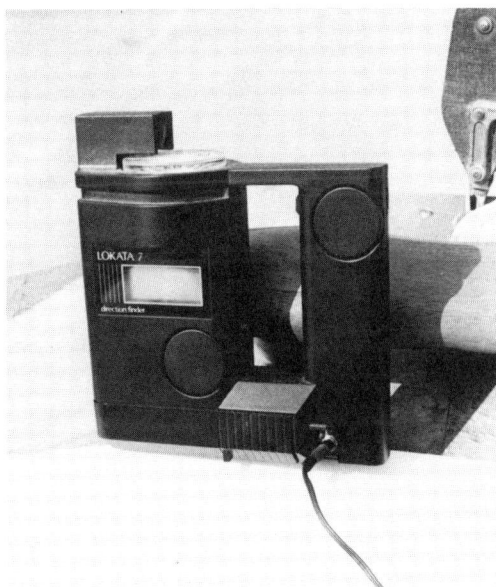

The invaluable RDF receiver.

The boat will follow a complex course without human guidance.

up a motor vessel with enough integrated electronic and other machinery to follow a zigzag course from one point to another and to make accurate turns at them with nobody on the wheel. The radar can be set with a guard zone to warn of obstructions and other ships, and then it will instruct the autopilot to avoid them. If the computer then predicts too little water in the approach channel, it will instruct the engine to slow down and wait for a more favourable entry time. There are, however, many inherent dangers in such total automation. They are obvious enough to set the Luddites screaming that it is totally unsafe to put your trust in machinery which can break down, and especially in electronic machinery which is still a developing science.

The answer is that it all depends on the machinery. There are few people who do not trust their digital watch which is electronic – I take mine regularly, together with my dive computer, out in rough weather and go subaqua diving to 30m (100ft) and more. Marine electronics live in a harsh environment, but most of them are as tough and reliable as the digital watch, and they are rapidly approaching the sophistication of those systems which land passenger aircraft daily with total precision.

Let me put all my cards on the table and admit that I am utterly devoted to electronic navigation and to other on-board gadgets, but am not yet convinced about complete automation and integration.

Even the dive computer is electronic.

Rather, I prefer to have the satisfaction of using them all myself as I take the boat from place to place, whilst letting one device verify calculations made with another, or having them supply a wealth of supplementary information to make my own decisions easier, more accurate and much safer. All electronic systems must be seen as tools; they are not a substitute for navigational skills, but an assistant to them. The electronics verify your pilotage and when they are properly used, one electronic system verifies decisions taken with another.

My own motor sailer is not too different from many other yachts, trawlers and small ships. Typically on a trip from our south coast home to, say, southern Spain, I will begin by plotting the course using traditional methods and will then verify them by reference to my Global Positioning System and my older Decca Radio Navigator. I have Loran-C as the backup for the areas further south along the route and generally set sail with both primary and secondary navigation systems running in tandem. My autopilot is controlled by an electronic compass to give greater accuracy and faster settling time. I also have a hand-held version for taking bearings and fixes.

The electronic depth indicator shows when I cross the shallow harbour bar, or when I reach the Hurd Deep. The speed log and digital timer are used for dead reckoning calculations. The electronic wind indicator advises the best course I can set to get the most out of a combination of low engine revs (shown by an electronic tachometer) and the sails.

Prior to departure, I will have received navigational warnings from the navtex and pulled the latest weatherfax prediction diagrams from the solid state FM

Electronic hand bearing compass.

Electronic log and wind indicator.

Much information from navtex.

Our portable radios are also scanners.

radio onto the laptop computer screen. This little PC also stores thousands of waypoint co-ordinates and allows me to predict how the tide will set me up and down the Channel. I can also use it to verify the direction of the 5-knot tide which I shall meet in the notorious Chenal du Four and the Raz de Sein. Once I have passed my route information to the coast-guard, my scanning VHF radio auto-matically monitors the necessary channels to keep me informed of shipping move-ments, navigational hazards and local weather conditions.

The radar is tuned to come on every ten minutes and to give me audible warning if there are any other vessels – or even lobster pot markers – inside the 4-mile safety circle which I have set. As I get closer to the coast I will use it to measure

Small yacht radars are very sophisticated.

heading and distance to buoys, light-houses and headlands. I shall also make these calculations with the other electronic wizardry and back this up with a battery-driven radio direction finder – a very comforting tool if all else fails. However, failure is not very likely. My massive battery banks are kept fully topped up by a TWC Adverc electronic regulator and each circuit is monitored by an LCD read-out. A single push of a button will tell me how much voltage is available to the line and how much amperage it is passing.

Meanwhile, the totally robotic autopilot holds my 7.5 tonnes on a steady course and tweaks the heading when I so ask it. Not having to steer with my gaze fixed on a compass card means that my eyes are much more free to keep a good look out.

I am dedicated to navigation by electronics, but above all *I still use my eyes.*

The battery state is monitored electronically.

## SUMMARY

- Electronic navigation is the perfect example of man's technological progression.

- The subject must be learned just like any other sea skill, and there is much to be learned.

- In spite of this need to learn the subject, electronic navigation is fun.

- Gadgets do not diminish the need for traditional skills.

- Electronic navigation is very efficient, but the best on-board aid will always be your eyes.

# 1
# BASIC NAVIGATION

Electronic navigation can only be made easy if the person using the very sophisticated machines (now available to even the smallest open boats) sees them as aids. Although they are very powerful they are relatively useless unless the user has enough knowledge of basic navigation to take full advantage of all their capabilities and functions. Almost all the new navigation and pilotage tools are controlled by some sort of electronic chip, or storage device, and thus are subject to the computer programmer's GIGO statement: Garbage in; garbage out. This means that you will not be able to derive maximum benefit from the super new systems unless you have sufficient basic navigational knowledge to drive them. At rock bottom you still need traditional, elementary chart and compass skills.

There is a plethora of electronic aids available.

A good compass is still the most important tool on board.

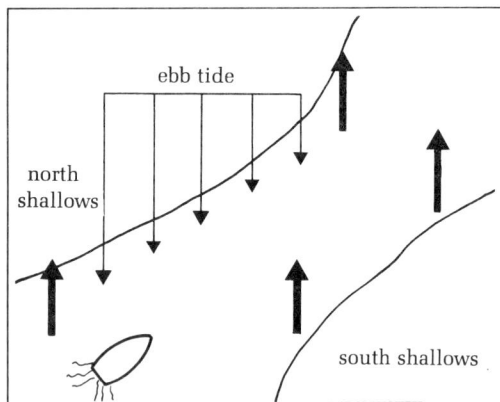

In this channel the autopilot is useless.

## Why Have the Old-Fashioned Basics?

There are some people who go to sea without having the basics of navigation, but I have also seen several mishaps to boats relying totally on Decca, video chart plotter, radar and interfaces to the autopilot.

These mishaps range from a man we pulled off the mud of the Exe Channel because he was relying on an automatic compass and autopilot setting (without realizing that the tide was crabbing him sideways), to the wreck of a £¼m motor cruiser on its maiden voyage in the hands of a first-time owner. She hit an isolated rock lying awash off Guernsey, whilst doing over 20 knots in control of radar with safety ring set and the autopilot guided by a Decca navigator. The new-comer skipper, who had refused the yard's offer of an experienced hand for this first passage, was reading a book at the time, safe in the knowledge that all the electronics had control of his vessel. He later admitted to being self-taught, and although there is nothing wrong with that he had not learnt the basics properly and was not sure about the basic chart symbols, and had no idea of tidal height and set. Clearly, he did not have enough basic navigational skills to realize that the electronic wizardry was getting the boat into trouble because it could not recognize a rock which was clearly marked on the chart. The machines were not to blame, however, and were only doing what they were told.

## The Marriage of Old and New

Equally, in sailing to northern Spain where this book was written, I had the safety, fun and satisfaction of passing between the south-west tip of Brittany and Ile Tevennec, through the notorious Raz de Sein on a day of moderate waves and dense fog.

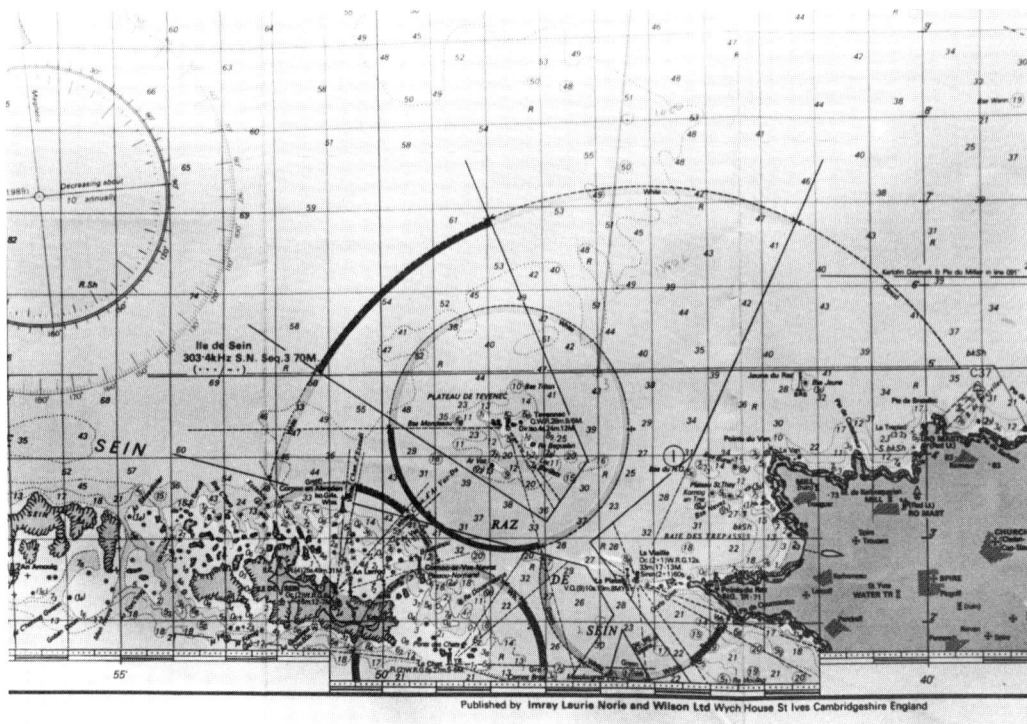

Published by Imray Laurie Norie and Wilson Ltd Wych House St Ives Cambridgeshire England

Ile Tevennec and the notorious Raz de Sein.

At other times I could not possibly have continued the passage, but with the radar spinning away and a GPS unit still, at that time, reliably accurate to about 20m (65ft) and backed up by the Decca, plus two depth sounders going in alignment with a large-scale chart, I knew exactly where I was at any given moment and where I would be ten minutes hence. I could also see the position and direction of all the other boats in the area.

This day was an excellent example of how a combination of ancient and modern can combine as a vast and comforting contribution to passage making and to safety at sea. I could not have managed without the electronics, but they could not have been driven without the basics.

## Devise Your Own Good Habits

My own system of passage planning and execution is not the only one possible, nor is it offered as perfection, but it does come from many sea journeys in motor sailers and in fast power boats.

Before I go to sea, all my routes are drawn on the chart and the turning places, or significant points along their lines (waypoints) are marked and given a name and a number. Waypoint latitude and longitude is noted on a passage log. Then with Breton Plotter, roller ruler and dividers, the course and distance between each is noted on both chart and log. Only when the work has been done manually do

The useful roller ruler.

I put the relevant information into the electronic systems. This allows the computers to check the work and vice versa, because very many electronic navigation errors are nothing more complicated than a person hitting a wrong digital key because of carelessness, or because the boat rolled at an inopportune moment.

By doing the plot both manually and electronically, any error in either immediately reveals itself. When it all goes well it is a great feeling. The converse occurs when you are in waves, fog and tide and your visual observation of where the boat is heading tells you that something is wrong with what the navigator is telling you to do, but you do not now have the time, nor enough free hands, to check what piece of false information, or which

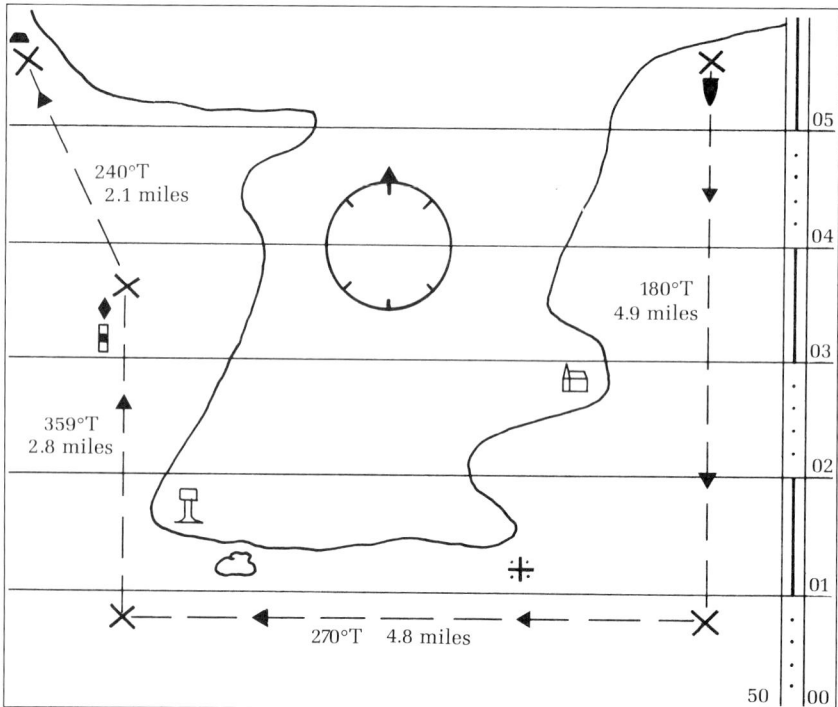

Draw the course on the chart.

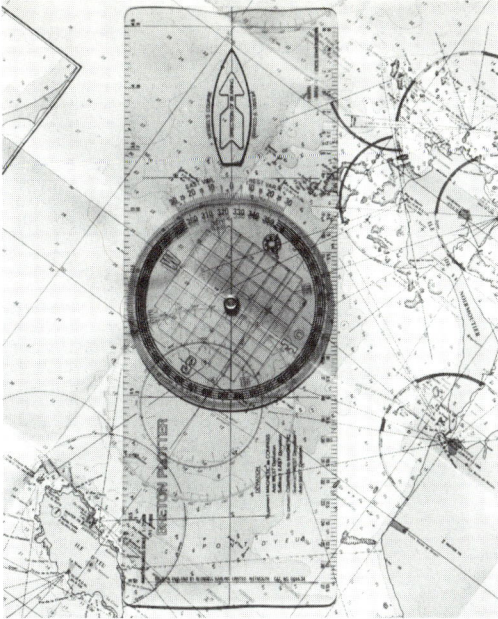

Check the details with the Breton Plotter.

wrong digit, is causing the problem. To be in such conditions and to not know precisely where you are or where you can safely steer, is no longer any fun.

The best advice – even if you have the most expensive electronic gear money can buy – is to do the planning in detail before

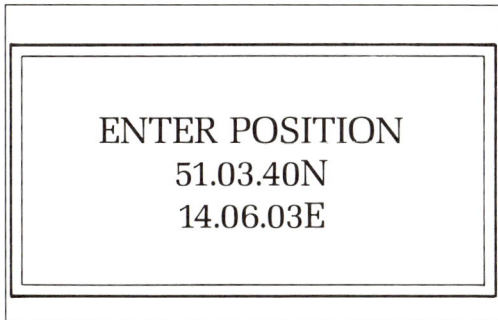

ENTER POSITION
51.03.40N
14.06.03E

A wrong entry puts the boat somewhere in the middle of France.

you set sail. Then you will fully understand what your equipment is doing and will be able to use it to the best advantage.

## Self-Taught or Navigation School?

The problem with many navigation classes is that the teachers often frighten off students by implying that there is something mystical or very complex about the skills involved – ergo they are very clever people because they can do it. Many of them also attempt to cram in more than the student can take and – in all honesty – often more than he will ever need to use. Most day-to-day leisure boat and professional coastal navigation is actually quite a simple matter of a few numbers and a bit of common sense.

With good equipment, a modern skipper has such rapid access to so many facts that he does not need to commit them to memory, and the in-built computers mean that all the number crunching, algebraic calculations are a thing of the past for most boat owners.

## Angular Distance

The most fundamental navigational function is, understandably, how nautical miles relate to degrees and how latitude and longitude (referred to as lat and long) are derived. The system obviously grew because man needed to locate and to relocate where he was at sea without having the land voyager's ability to build cairns and other stone route-marking piles or to relate to mountains.

Angular distance, degrees and miles become simple to understand if you

15

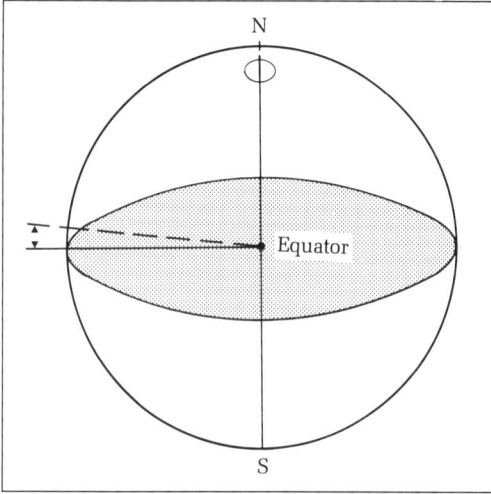

Imagine a bar from the earth's centre to the Equator. Raise the pivoted bar to get 60 miles for each degree.

imagine a solid bar, pivoted in the very centre of the earth's sphere and stretching out to the Equator. At rest it would be horizontal and the North Pole would be vertically above the pivot. To get from the Equator to the North Pole, the bar would pass through 90 degrees. By calculation, we know that raising the bar by 1 degree causes its free end to move 60 nautical miles north of the Equator. Come up 2 degrees and the Equator is 120 nautical miles behind you, and so on. This information is conveyed on marine charts by various clever projections. These are necessary to cope with the fact that the earth is not quite a complete sphere, and whereas lines drawn at equal distance north and south of the Equator always

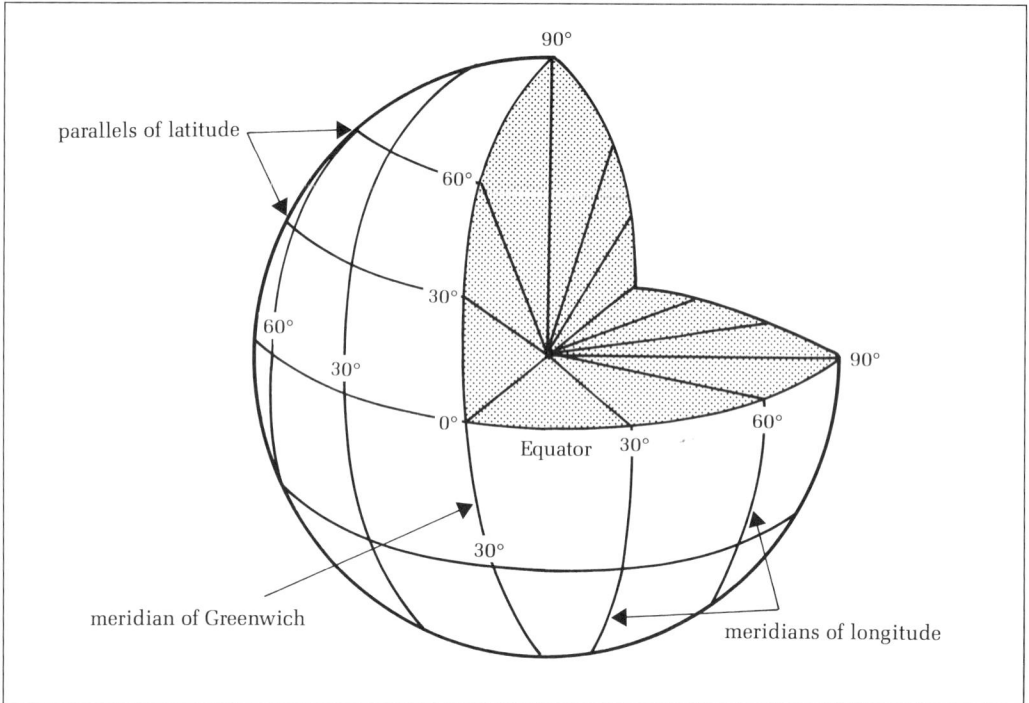

Parallels of latitude and meridians of longitude. Parallels remain parallel and meridians converge.

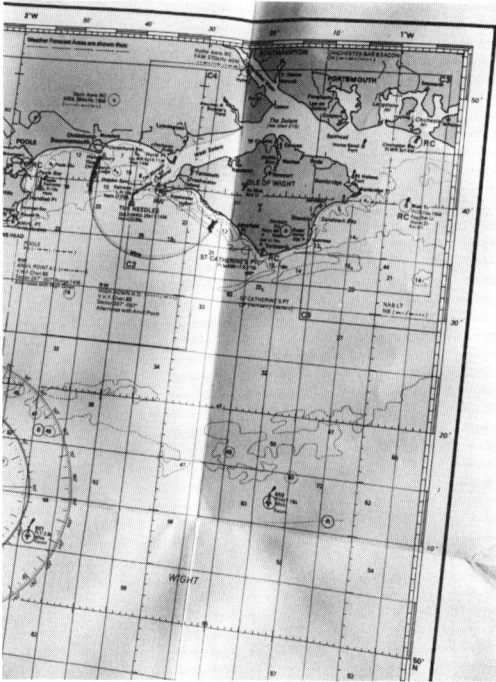

The chart's latitude scale measures nautical miles.

remain parallel to each other, vertical lines of longitude converge on each other as they approach the poles.

This is a very simplistic outline, but it explains why nautical miles on the chart are always measured by reference to the lines of latitude on the left and right edges. Because the cartographer has to adjust the length of these lines on paper to cope with the problems of geography and of representing a curved earth on a flat print, you must only take distance readings from the vertical edges and at a point roughly level to where you are navigating on the chart. The point where a line of latitude crosses a line of longitude always gives you a statement of position, but nautical miles are only measured by reference to latitude – the horizontal lines on the chart.

## The Nautical Mile

To date, the words nautical and mile have always cropped up together in the text as a

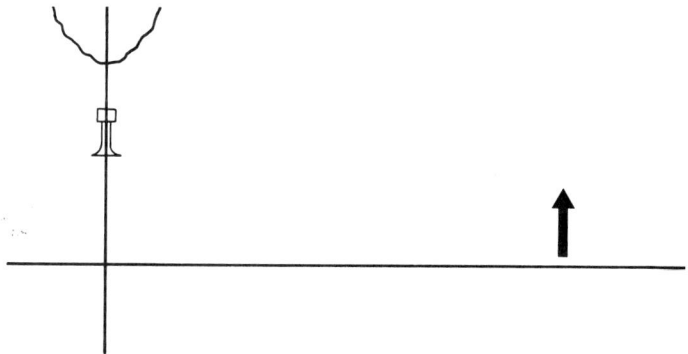

'You are due South of the lighthouse and due West of the spar buoy, Sir.'

Where two lines cross, we have a position fix.

The Land Mile is usually worked in quarters.

The Nautical Mile is longer and nowadays worked in tenths.

Statutory mile and nautical mile compared.

measurement of distance. The facts above show why the nautical mile is a precise distance determined by geography, whereas the land, or statute mile, was arrived at by royal whim.

A sea mile is (almost exactly) 1,870m (6,080ft, or 2,026yd). Most navigators do their mental arithmetic in approximations and think of the nautical mile (10 cables in old parlance) as a 200m (220yd) distance. This is obviously not spot-on accuracy, but

Measuring a distance with dividers.

is perfectly adequate for practical pilotage. Its main virtue is in making all arithmetic very much simpler and it is perfectly good for most speed and distance calculations. Throughout this book, all miles are nautical miles unless otherwise stated.

## Seconds or Hundredths?

Because a nautical mile is one-sixtieth of a degree, the vertical chart distance between two points located one mile apart is still occasionally referred to as 'one minute of arc'. In some older pilotage books, fractions of a sea mile are still calculated in 'seconds', but a more modern practice (largely brought about by electronics and computers) is to state these smaller divisions in decimal notation, in other words tenths and hundredths.

Thus, a safe entry point to my own home port is written as:

[50:43.10N    02:55.90W]

If I give this over the air, it becomes, 'Fifty degrees, forty-three minutes decimal one zero north: zero two degrees, fifty-five minutes decimal nine zero west'.

This use of tens and decimals makes the

```
┌─────────────────────────────────────┐
│  DECCA PRO                            │
│                                       │
│  ┌─────────┐   ○   ○   ○   ○         │
│  │ EST POS │ ▼                        │
│  │         │       ○   ○   ○         │
│  │ 10·24·18N│  ○   ○   ○   ○         │
│  │ 03·17·15W│ ▲                       │
│  └─────────┘       ○   ○   ○         │
└─────────────────────────────────────┘
```

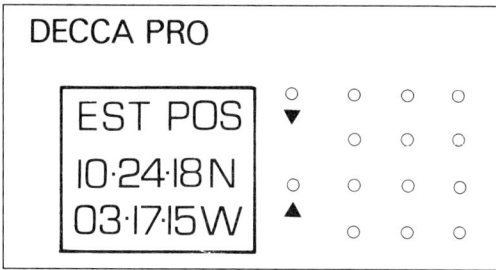

Decca reads to tenths then hundredths.

rounding-off approximations described above (2,000m and so on) even more useful for rapid pilotage and quick checks. It means that the first figure after the decimal point of a Decca read-out is worth roughly 200m (220yd) and that the second decimal place is good for 20m (20yd) or so. Even with normal sophisticated electronics, if anyone tells you that he can work to closer tolerances than this, he should be treated with a large dose of doubt.

By now, you should be getting the picture that modern navigation is a mixture of slow, painstaking calculation, rapid and very precise measurement by computers and a bit of guesswork thrown in – a hodgepodge which is very practical and which works.

## The Knot

The unit of marine velocity is the knot. A boat travelling at 5 knots covers a distance of 5 nautical miles in one hour, so there is no such phrase as 5 knots an hour.

## Tides and Currents

No skipper can get from A to B efficiently without a knowledge of tides and tidal

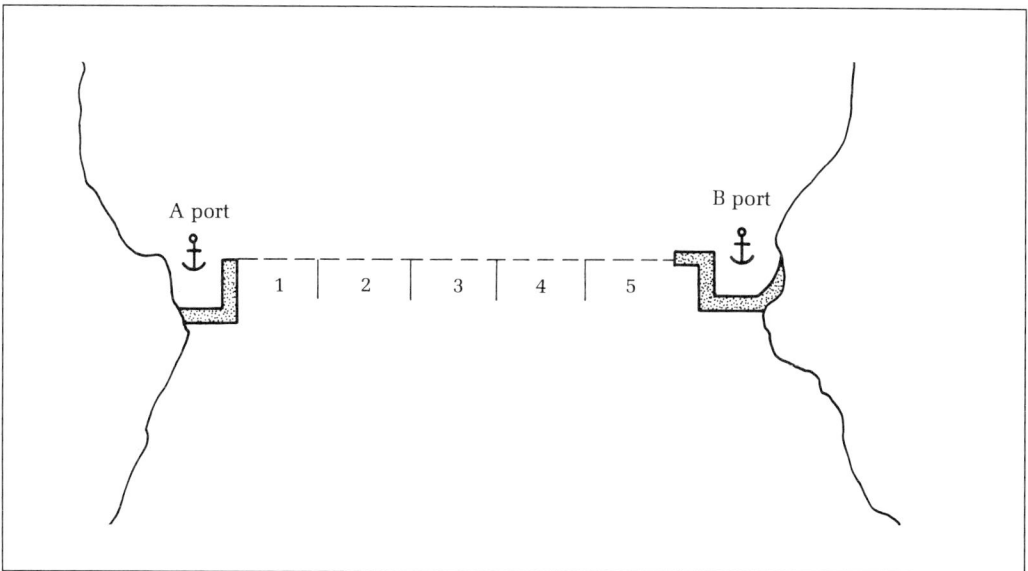

Distance from A port to B port is 5 miles. To get there in an hour, you steam at 5 knots.

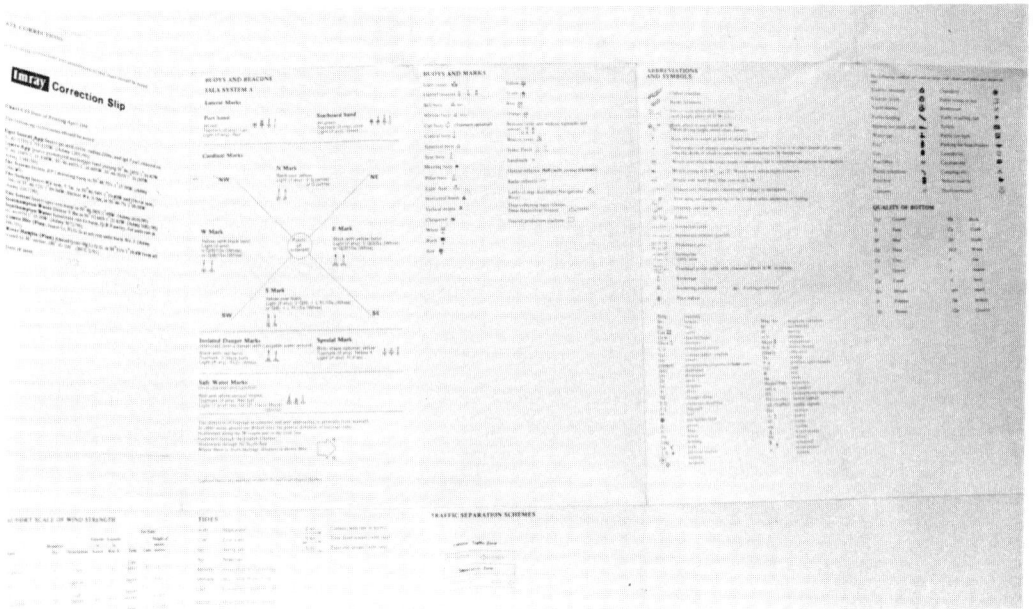

Imray charts show all the symbols on the obverse.

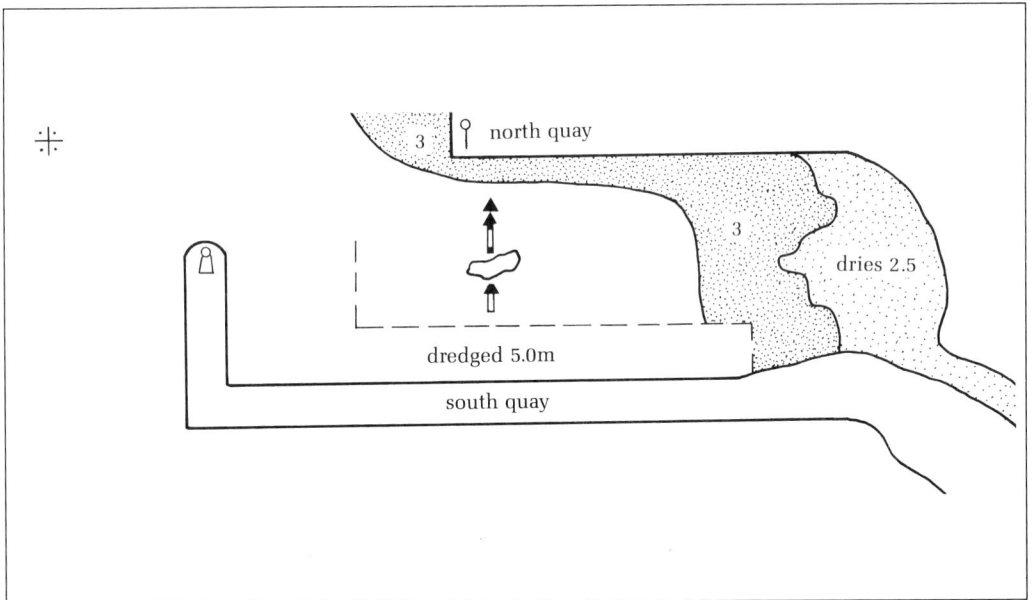

Rock and harbour depths are shown by symbols.

streams – those currents which largely run parallel to the coast and change direction as dictated by the ebb and flow.

The influence of tides can be seen all over the chart and it is rarely far from the passage-maker's mind. Charts contain information on where depths can be expected below 20m (66ft), 5m (16ft) and so on. They also show depth over rocks and the height a rock or sand bar will be above sea level when it dries out.

## Lowest Astronomical Tide

All the chart measurements must obviously relate to the same water-level and are generally reduced to Lowest Astronomical Tide (LAT) which is the lowest point to which the sea's level will be reduced at the lowest tide which can normally be expected.

Thus, a rock shown with the symbol 4.9 will always have 4m and 90cm (16ft) of

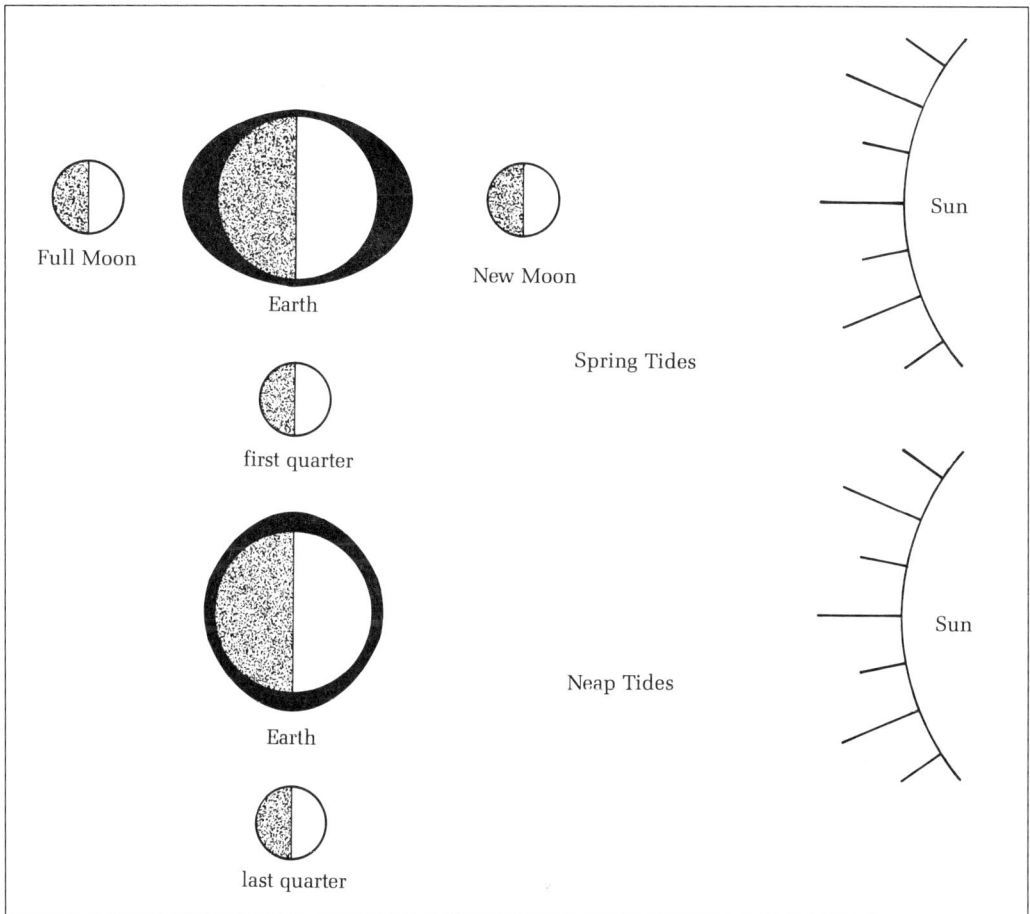

Sun and Moon in line means big tides (top). Sun and Moon at an angle means small tides (bottom).

water over it, at the very minimum, even on a day when a huge spring tide has drained the maximum amount of water away from the area. Similarly, a harbour shown to dry out to 1.5 will be 1.5m (5ft) above that sea level at 'dead low water springs'. At other times it will be standing less proud and will be just awash when the water-level has risen 1.5m (5ft) above this 'lowest ever expected' point.

## Tidal Range

The depth figures quoted by tide tables give the tidal range. The low water figure shows how much water can be expected above chart datum or LAT at that time, and the high water figure gives similar information about 'the top of the tide'.

The difference between the two figures is the tidal range. A big range means that much more water must be moved in and out in the same time as a small range, so the currents around headlands and pouring down the Channel will be much stronger.

## Using Approximations

There are enough complications at sea without confusing your brain with too much precise arithmetic. Seasoned navigators will always work to the nearest easy figure – generally 25cm (10in). So a low water of 1.4 will be noted as 1.5 and a high water of 6.9 will be referred to as a 'seven metre tide'.

Similarly, do not confuse yourself about time. A high water at 18.55 is a 'seven o'clock tide' in anybody's real language.

## The Rule of Twelfths

You can ascertain the depth of water at a given place by a variety of means from tidal interpretation curves to electronic

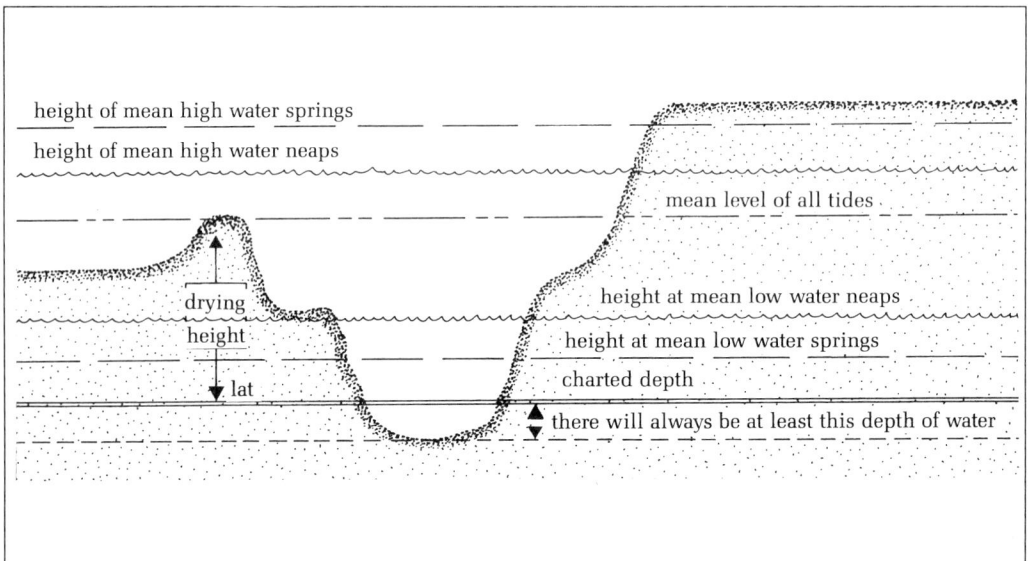

Tidal flow and the tide's range found from charts and almanacs.

---

**Flood tide rise according to Rule of Twelfths will be:**

| | | | |
|---|---|---|---|
| first hour | one-twelfth | fourth hour | three-twelfths |
| second hour | two-twelfths | fifth hour | two-twelfths |
| third hour | three-twelfths | sixth hour | one-twelfth |

---

calculators, dedicated electronic tide predictors and computer programs. However, a cheap and simple way to get the same information in a digestible form, which is approximate but accurate enough for practical navigation, is to use the Rule of Twelfths. This dictates that the tide will rise or fall one-twelfth of its range in the first hour of its period, two-twelfths in the second hour, three-twelfths in the third hour and so on as shown in the table above.

By using a calculator, this can soon be transcribed on to a table noted on the ship's. log. If we have a noon low water with a range of 6m (20ft) for the day, the tide would behave as described in the example overleaf. It begins with the tide rising one-twelfth of its range in the first hour from 12.00 to 13.00.

*En route*, the skipper can look at a channel whose charted depth shows 1.0m (3.3ft) and know that if he is going through at

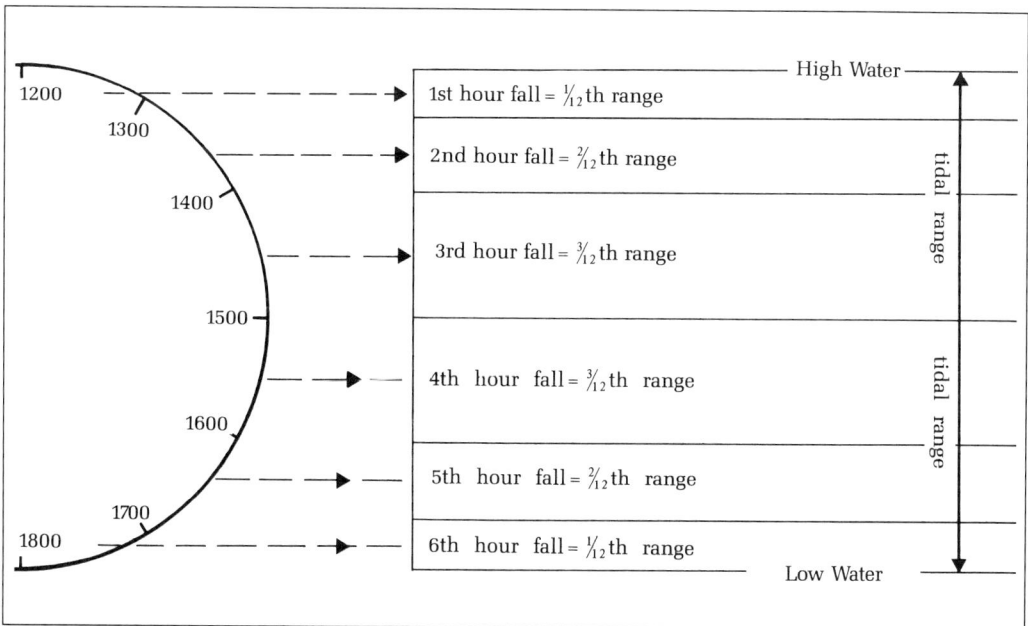

The Rule of Twelfths for one tidal cycle.

| Time | Fraction | Rise | Depth |
|------|----------|------|-------|
| 13.00 | one-twelfth | 0.5m (1.6ft) | 0.5m (1.6ft) |
| 14.00 | two-twelfths | 1.0m (3.3ft) | 1.5m (5ft) |
| 15.00 | three-twelfths | 1.5m (5ft) | 3.0m (10ft) |
| 16.00 | three-twelfths | 1.5m (5ft) | 4.5m (15ft) |
| 17.00 | two-twelfths | 1.0m (3.3ft) | 5.5m (18ft) |
| 18.00 | one-twelfth | 0.5m (1.6ft) | 6.0m (20ft) |

15.00 he will have 4.0m (13ft) of water. If this same channel is shown to dry out to 1.5m (5ft), it will be awash at 14.00 and have 3.0m (10ft) depth at 16.00.

The same simple, practical calculation can be applied to the ebb. It is recognized as an approximation, but the same can be applied to much tidal predication information and cruising navigation generally.

## Magnetic Variation

Ever since the early Chinese discovered that lodestone possesses the peculiar property of always aligning itself in a particular direction, inexperienced navigators have been wrestling with magnetic variation. This phenomenon occurs because the precise location of the North Pole moves slightly each year. Its annual shift is only a few fractions of a minute of arc (seconds) and the amount is noted on all chart reprints.

Again, many navigation courses and books are guilty of overpreparing students and simultaneously making magnetic variation too complicated. For example, the direction of this variation changes according to the hemisphere, but how many British pleasure boat sailors will ever cross the Equator, or go to America in their boats? Thus it makes more sense to retain a mental note of the simple, single figure addition or subtraction required for the area where you normally sail. Learn it once and use it all the time.

In actual, on-board practice, especially allied with good electronic navigation aids, these minor shifts of magnetic north make very little difference. Most navigation systems give their directions in true north, which never changes, and the present magnetic variation is then added to give what can loosely be called the 'compass course', or the figure which you will set on the steering compass lubber line. Thus if your radio navigator tells you to steer 270 and the local variation is 5°W (this is a typical UK figure), you set your boat compass to 275 and steer that course. To avoid any confusion at the steering position, we always work in degrees T and leave whoever is steering the boat to make the actual adjustment to degrees M, in the hope that this constant practice removes confusion and reduces the areas of possible mistake.

The situation is made more rule of thumb because of three degrading factors:

1. Very few boat compasses are accurate to one or two degrees on all headings.
2. It is possible to rotate the boat and to note how far out the compass is on many different headings compared with known

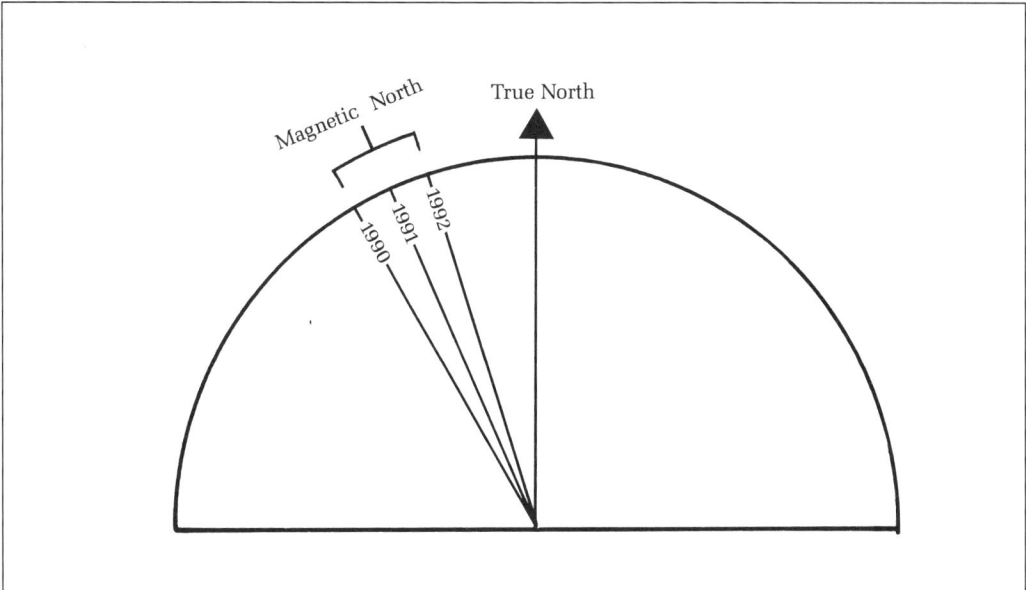

Magnetic north moves a little each year.

onshore transit alignments (make a deviation card to keep near the tiller), but this is physically so difficult to arrange and so time-consuming to do, that very few boat owners even think about it – we just accept that our steering might be a couple of degrees out.

3. Our conscience is eased in this matter because anybody who tells you that he can hold a small boat permanently within a few degrees of his intended course, even in a very moderate sea, has probably spent more time at the bar than on the wheel. Even Wilbur, my very expensive and very accurate electronic autopilot, does not claim total accuracy to one degree here, not only because of the way in which all boats yaw from side to side, but also because boats are mostly being affected into zigzag steering lines, or made to slip sideways by the actions of wind, waves and currents.

All this does not mean that the current leisure navigator is sloppy or inaccurate – quite the reverse in fact. He realizes that these imperfections exist and he uses the super modern aids constantly to make one system check out another and to use their consensus to make his judgements. Put a different way, if he was to rely on just one navigation instrument, for example, the sextant, he would be lost for much of the time, but by using everything at his disposal, he generally has a very good idea of exactly where he is and where he is going. So, thanks to electronics, our modern navigator is a very accurate operator, no matter what the season or weather. He also enjoys the fact that navigational practice is much easier than its theory; that alone makes it safer.

This is also the argument against all those boring, traditional ancient mariners who decry modern technical aids with the whinge 'What happens when it breaks down? Then you are in trouble.' If it breaks down, it can probably be repaired, whilst I go on to the backup system on its own, separate circuit. If the electronic compass goes on the blink, I still have several magnetic units on board. Good seafaring is a belt and braces pursuit of plenty of knowledge, some DIY mechanics and lots of backup. In this sense, we are no different from those highly advanced, ultra-progressive mariners who first abandoned the pelorus in favour of the sextant. They must have faced the pessimists asking what they would do if they dropped it and pushed the mirror out of alignment.

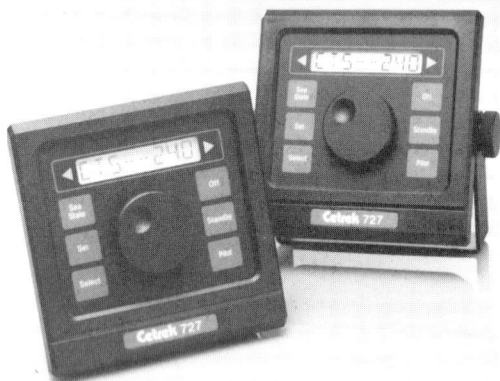

Wilbur, our very accurate autopilot.

Progress is safer and more fun than sticking entirely to traditional methods. We who go to sea, even in small ships, live in very exciting times, so if you can afford modern technical aids, enjoy them and be safer because you have them.

---

**SUMMARY**

- All navigation systems need operator care: garbage in, garbage out.

- Every boat and every skipper uses a different variation of the basic themes. Through good practice you will discover what works best for you.

- Electronic navigation always requires 'input' which is then checked by manual and literary reference methods.

- Angular distance means that 1 degree equals 60 nautical miles.

- Do not try to be too clever; two decimal places of a minute of arc is accurate enough.

- One nautical mile per hour is called 1 knot.

- Lowest Astronomical Tide (LAT) is the lowest level the tide will ever reach. These levels are the depth figures on nautical charts.

- Tides can best be estimated on the Rule of Twelfths.

- Always work your ship in degrees true then adjust for magnetic variation at the steering position.

- Your electronic navigation will only ever be as good as your basic skills.

# 2
# RADIO NAVIGATION SYSTEMS

Next to a good compass, a reliable navigation system would come very high on the equipment shopping list of most skippers. Happily, the mariner is currently very well blessed with external methods for fixing his position and providing him with much more useful information besides. Primarily, he can choose between a land-based system using radio waves, or a satellite-controlled system using fast and accurate clocks.

The future of electronic navigation probably rests with satellites, but radio navigation will still be with us for some time to come, with a number of countries already committed to its retention in full working order well into the twenty-first century.

Perfection; compass and two navigational systems.

## What is a Radio Navigator?

Currently, there are two important and publicly available radio nav systems in operation. Decca is the one more widely used in Northern Europe and in certain parts of the Mediterranean. In other areas Loran-C is more common and it is the only system available in quite a few areas of France, Canada and America, the latter being its country of origin. Both Decca and Loran-C are well developed, well known and relatively free of major problems. The shipboard antennas and receivers are within the budget of most owners and their toughness and easy installation make them suitable for virtually every craft from mega yachts down to subaqua club inflatables.

The two radio navigation systems have

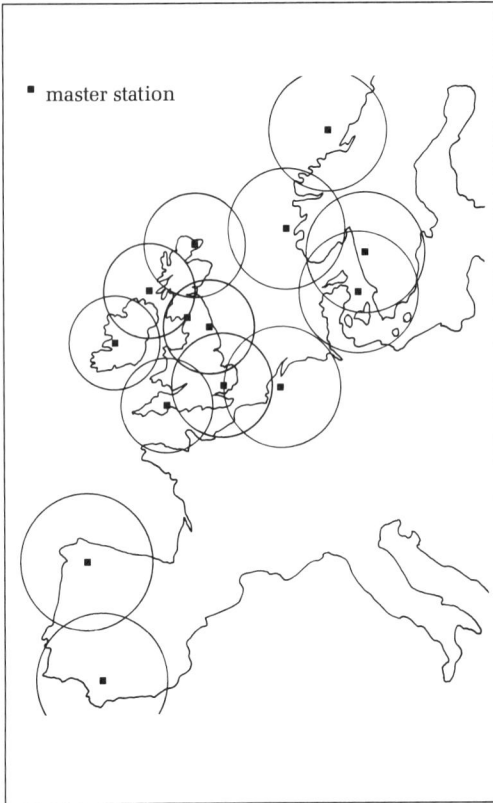

master station

Decca coverage is good for northern Europe.

served by twenty-four transmitter chains, each of which can be used (with varying degrees of accuracy) by any vessel located within approximately 250 miles of the master station. The master station is the most powerful of the four transmitters making up a chain. The others, called slaves, are set roughly in a triangle around the master, at distances of 50–150 miles. The slaves are designated by the colours red, green and purple. All four stations transmit precisely measured and monitored radio signals twenty-four hours a day.

The simplest way to get a picture of how Decca works is to view the master station as the place where a pebble has been dropped into a very calm pond. The ripples spread out concentrically from the

You can spot the Decca boat by its antenna.

more similarities than differences. Decca uses a slightly higher frequency band than Loran-C and in consequence has less range. Because Decca is still the more current fashion in Europe, the ideas and explanations which follow are based on European Decca, but will also almost totally apply to Loran-C receivers, including how they function and how they can best be used on board.

## The Decca Chains

Europe is a typical Decca example. It is

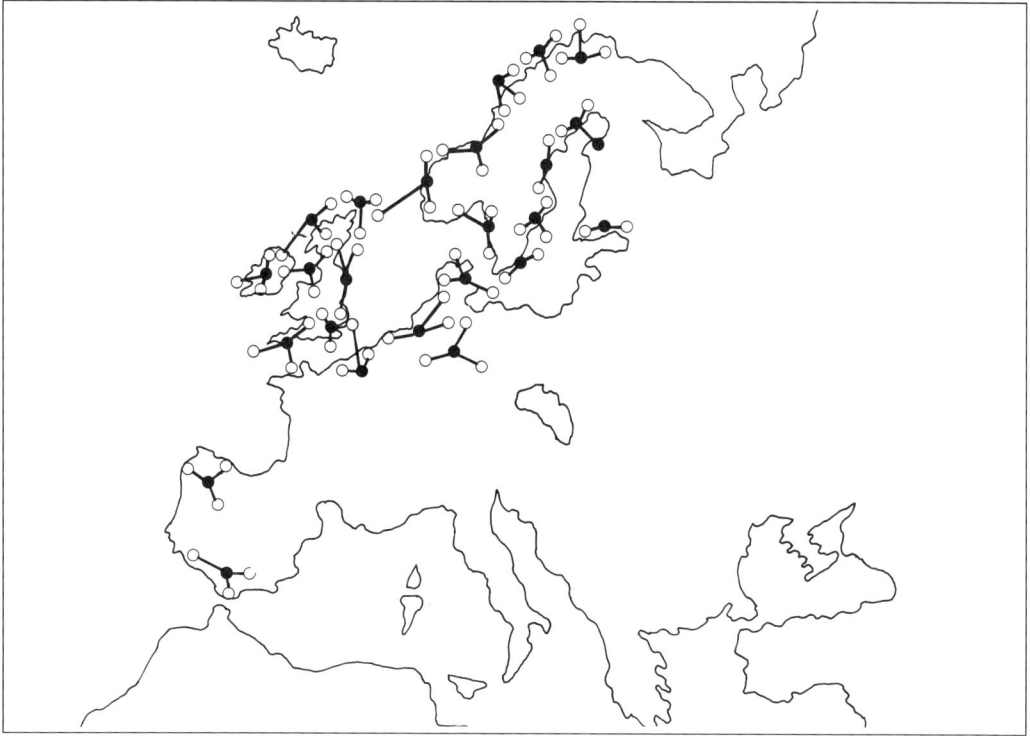

European Decca chain arrangement.

drop point with a uniform distance between them, and their speed can be measured. Thus, if a boat is passed by one of the wave ripples travelling at known speed, we could calculate how far away the boat is from the master station, or pebble drop point.

We would know the distance from the master, but not the direction. So, we drop in another pebble at a different place. The waves from this second, or slave sender, can also be measured for speed and distance. Where any two of the waves cross, we should have an approximate position, verified by reference to two stations. It would, however, not be very precise, but located somewhere in a long thin slice of

space, as shown overleaf. The solution is to drop in a third pebble to give our master a second slave and three wave crossing places. We now already have the navigator's classic, small cocked hat triangulation: a precise position which can be even further refined by the arrival of a wave from slave three.

The velocity of radio waves is known to be the same as the speed of light and the distance which they travel per millisecond can be measured. On the boat, the Decca navigator is a combination of radio receivers capturing the signals and a small computer working out the distance from each transmitter to establish a very precise position.

Drop a pebble in the pond.

Where two waves cross, we have a fix.

Three crossings give a cocked hat.

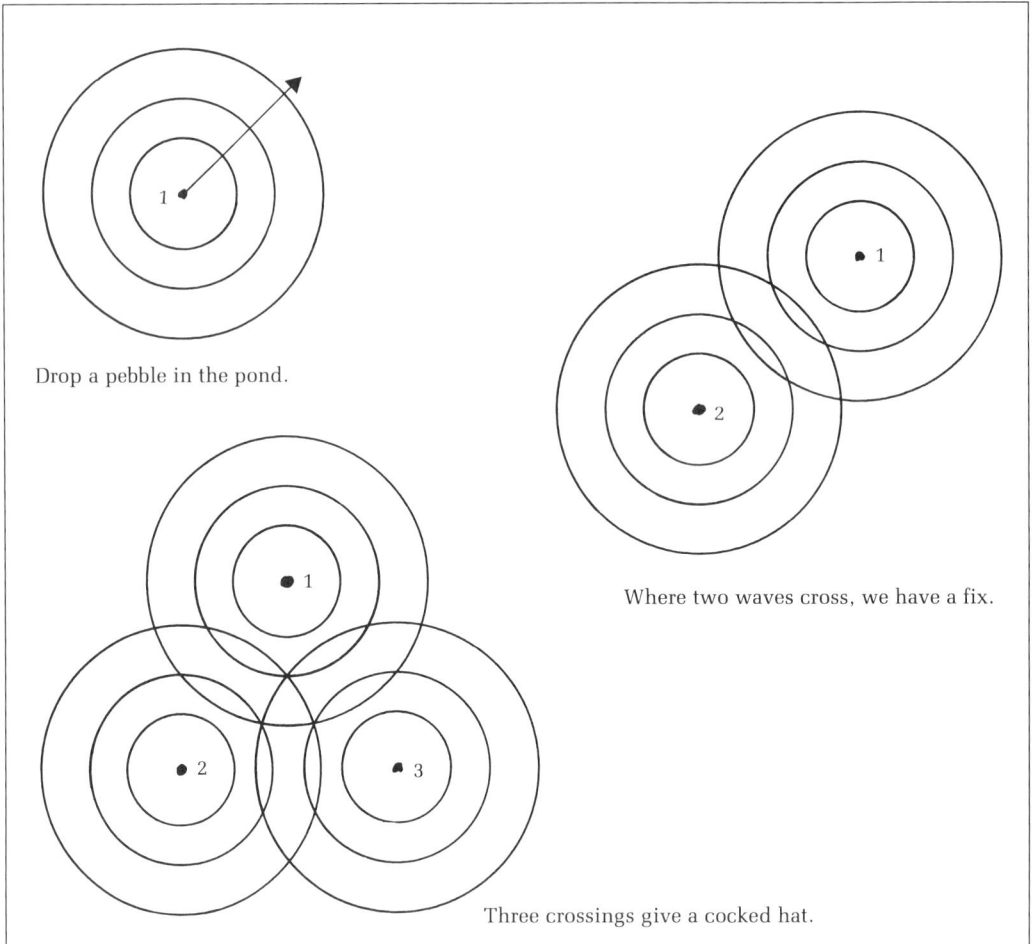

This position can be converted into Decca LOPS (Lines of Position) which are still used by many professional fishermen, but which needs a special chart with a coloured overlay, or into lat/long, which can be interpolated on to a standard, less expensive chart.

Once the computer has this initial position in its memory, it can calculate quickly the distance and direction to any other position which the operator gives it. This computing power is a major part of the radio electronic navigation system's sophistication and explains its ability to perform some very useful pilotage functions.

## Theory Versus Practice

Most Decca sets have a display which reads out to two decimal places of a nautical mile – say 60ft, 20yd or 18.7m, and there are times when conditions are

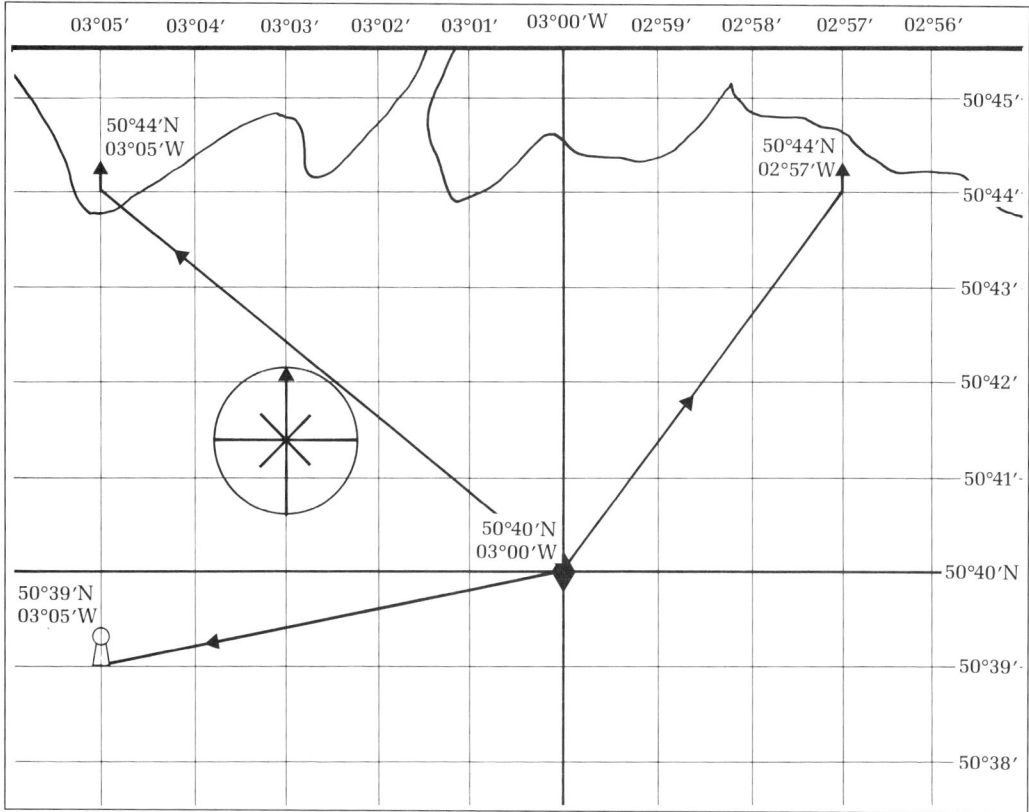

The computer can work out course and distance to any other place if its latitude/longitude is known.

such that this accuracy is achieved. Unfortunately, we do not live in a perfect world and a number of factors combine to reduce this theoretical precision. Decca technicians recognize these imperfections and have devised the means by which some allowance can be made for them. By thousands of observations in many areas and under all sorts of seasonal and climatic conditions, a huge database of possible errors has been built up. This can actually be built into the receiver or, as with most Decca receivers, translated into a function which displays the circle of

possible error. This varies from place to place, but its inclusion does at least give the skipper some idea of the accuracy of the positional information on the screen.

In my home area of Lyme Bay, the circle's diameter is only 0.06 miles, but where I am writing here in north Spain it is 0.47 miles – this can reach to over a mile in other places. This means that at home, my position might be anywhere within a circle about 100m (110yd) across, but in Galicia the circle is almost 1 mile.

The size of the circle is dictated by such things as distance from the master station,

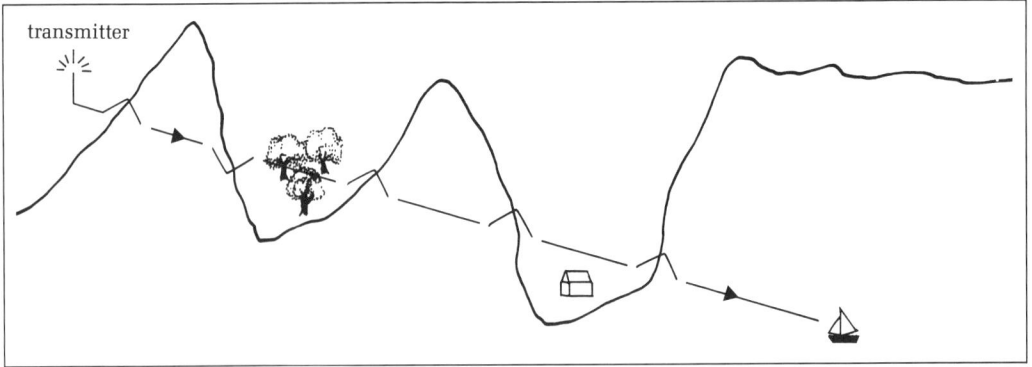

Land/sea boundaries distort radio waves.

atmospheric conditions, and whether the radio waves cross any phase and speed altering land or sea boundaries. They are also affected by the time of day, which in turn alters the altitude of the ionosphere, the electrically charged layer of the atmosphere which reflects radio waves. This change of height alters the angle of reflection and causes the waves to arrive at the aerial by a peculiar route and not at the same speed as signals from the other stations of the chain. They will also be slightly out of phase with each other, which makes it difficult for the integral computer to resolve them into pinpoint accuracy. Engineers take the most pessimistic view. The quoted possible error is the absolute worst that can be expected. Most experienced Decca users see the figures as a note of caution and a reason to double check, but will agree that their on-board navigators perform about 90 per

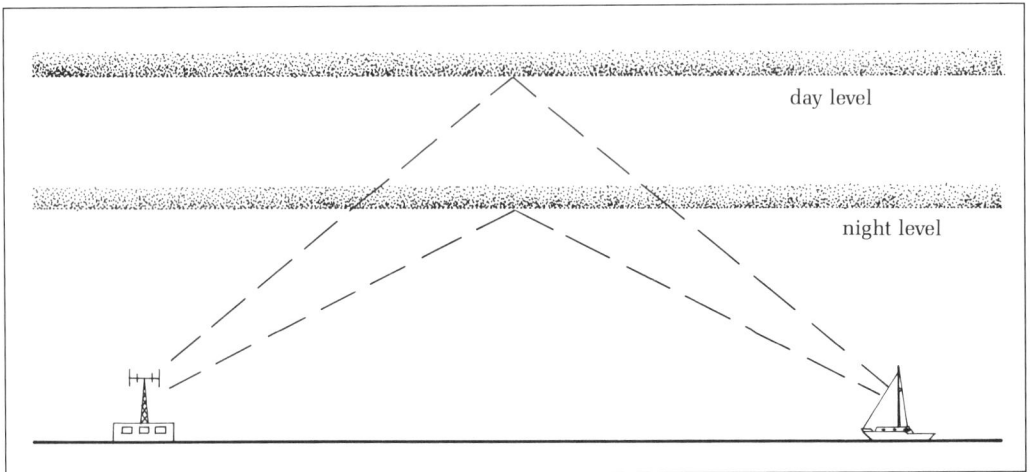

The changing ionosphere alters the wave's time of arrival.

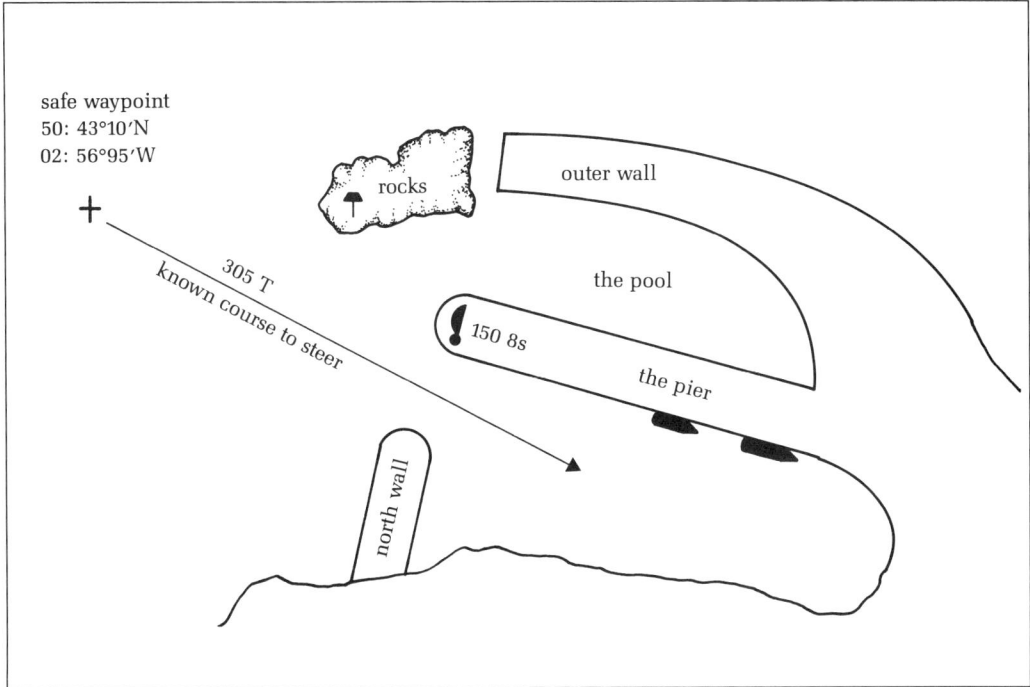

safe waypoint
50: 43°10′N
02: 56°95′W

rocks

outer wall

the pool

305 T
known course to steer

150 8s

the pier

north wall

Decca brings us back to a safe harbour entry every time we come home.

cent better than this for about 90 per cent of the time.

Decca is an excellent system. On a clear, summer day along the coast of a good signal reception area the accuracy is very impressive – really down to a few metres. It also has tremendous repeatability. By this I mean that if you park alongside a navigation mark, or over a fishing wreck, or at a safe place to enter harbour and note down the precise lat/long as shown by your particular Decca set, the system will always guide you back to within a few yards of it. If the atmospheric pressure, weather and time of day are identical to when you took the readings, you can totally trust the system to these few metres. It should, however, be understood that your neighbour's rig might not show

exactly the same figures and that the second decimal place might not always be the same. But who reckons to be able to navigate to two decimal places?

Possibly the best analogy for Decca's accuracy comes from an angling charter boat skipper who uses his Decca daily to put his clients over offshore wrecks. 'My compass and autopilot will find the football pitch for me, then the Decca puts me in the right part of the penalty area, but I sometimes – not always – need to let the other electronic tools sniff about in a tiny circle to find the actual penalty spot.'

My own experience confirms this in another way. One night I switched my Decca off when I was tied to a pontoon in Cherbourg Marina. When I fired it up again the next morning, the pontoon had ostensibly moved about 30m (100ft) north and 60m (200ft) east. However, I also went to Cherbourg at the beginning of September in three successive years and each time used the same lat/long reference for the West Entrance. The Decca guided me to within a few metres of where I wanted to go on each occasion. I can therefore always use it with confidence and a touch of seaman's prudence.

## Where Am I? Where Will I Be?

A couple of decades ago just to know your position and to be able to transcribe it on to a chart within seconds would have seemed like a miracle, but our present tiny black box receiver and computers can do plenty more. Once the machine is getting a continuous update of position and how this position is changing from one minute to the next, it can begin to compute lines of actual travel and even to compare them with where you hope to go.

## A Note on Installation

Before taking you into a practical radio navigation session, it is worth pointing out that both Decca and Loran are very easy to install. Once the bracket to hold the display has been mounted at a safe distance from the VHF radio and your magnetic compass, the task requires no more than the connection of two wires to supply the current, and possibly a third to earth the system to a grounding plate or other metal surface in connection with the water. Earthing, or grounding, is very important on vessels driven by outboard motors, as these are infamous for the

creation of electrical noise. If your Decca is showing poor results on such a boat, you will be offered all sorts of engine cowl shielding devices and compounds, but often the simplest solution is to install an earthing plate well forward and away from the motor and to ground your radio navigator through it.

Installation of the antenna is also relatively simple. If it can be clear of the electrical field which is generally built up around a petrol engine, so much the better. There is, however, nothing much to be gained from altitude. An antenna mounted on a guard rail will perform as well as one on the mast and will have the other advantages of low wind interference and being able to keep well clear of the VHF aerial, whose emissions will almost always wipe out any Decca signal to an antenna placed within a metre of it.

## Final Approach

Mariners are forever in a state of debate about which navigation system to install, and even debate if a terrestrial radio system is worth having in view of current and expected satellite derived coverage. The advice will depend on your cruising area and your pocket. If you rarely go out of sight of land, or sometimes cross the Channel but never indulge in 'blue water' cruising, you will be very adequately suited by Decca and Loran – depending upon where in the world you live.

However, because being at sea is always safer when you have a backup for absolutely everything, if your wallet is sufficiently large, equip the boat with both radio and satellite navigators running from different electrical circuits. The one will then check out the other and you can

Our boat's earthing anode obviously works well.

have them running on different displays, which saves a lot of button pushing.

Besides, boating – and especially electronically assisted boating – should be fun, so if you have the wherewithal to increase your safety and your fun, you will be well advised to use it.

## SUMMARY

- Decca is the more usual North European position-fixing station.

- Decca relies on a master station and three slaves: red, green and purple.

- Radio waves travel at the speed of light.

- Decca's circle of possible inexactness varies from area to area.

- Decca loses precision at night, in winter and in very poor weather.

- DIY radio navigator installation needs care – especially with reference to a steady voltage and proper earthing.

- Loran-C is the US radio position-fixing system.

- Loran-C has all Decca's virtues and vices; the two are very similar.

- You can best use Loran north of the Humber and south of Coruna.

- There are low-cost systems available for coast hoppers.

# 3
# SATELLITE
# NAVIGATION

The Global Positioning System (GPS) is the most important navigational development of the twentieth century. It would be difficult to overestimate the effects which it will have, not only on how we find our way about, but on who will be navigators. GPS is not merely a sea tool, but can be used in cars and in addition to being fast enough to be usable in modern fighter aircraft, it can also display its altitude.

Versatile GPS can even be hand-held.

Satellite derived position fixing is not new, but all the older systems pale into absolute insignificance when compared with the power, precise accuracy and total global reliability of a 100 per cent operational GPS set-up. It has cost the US Department of Defense many millions of dollars to put satellites into working orbits, but the spin-off for the rest of us, plus the amount of money various industries will make from it, will give good return on the investment over a long period of time.

As its name implies, GPS signals can be received everywhere in the world and they are largely unaffected by season, weather, geographical factors and all the other attenuating conditions which weaken Decca, Loran-C and so on.

## Satellite Ranging

The technical and most explanatory term for how GPS works is 'satellite ranging'. This means that the on-board equipment selects several satellites from the twenty-one in orbit about 11,000 miles out in space (plus two reserves) and calculates the range, or distance, to each of them. Because there are so many units making a 'birdcage' around the Earth, there are

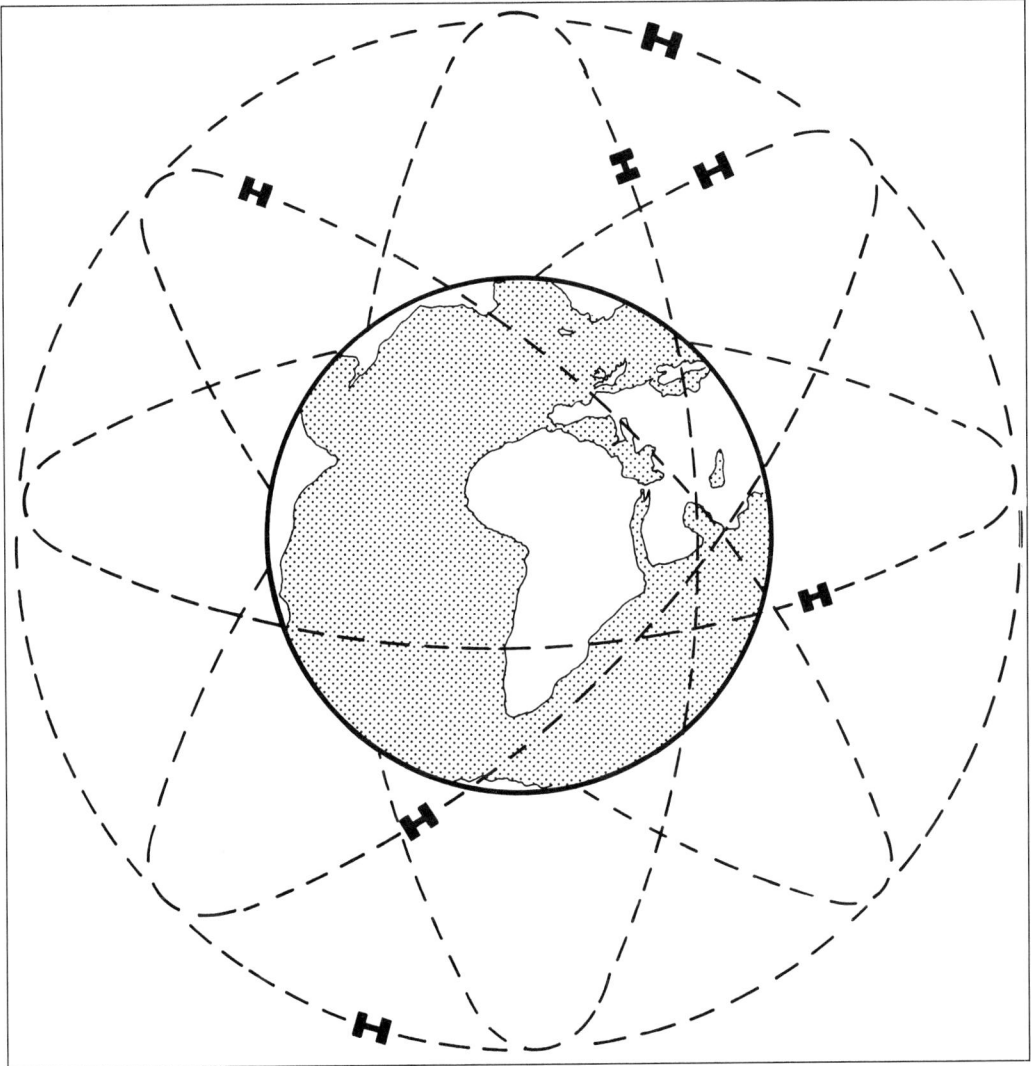

The GPS satellite birdcage gives total cover.

almost always enough above the horizon to give sufficient data for position fixing. This choice is part of the system's strength. There are some Decca and Loran areas where the position lines (arcs) touch each other over a large amount of space because the angles from boat to trans- mitters is not good. By picking the best satellite geometry, a GPS receiver does much better than this.

The satellites themselves are sur- prisingly small, measuring only about 5m (17ft) across the power/antenna panels. Unlike television satellites, they are not

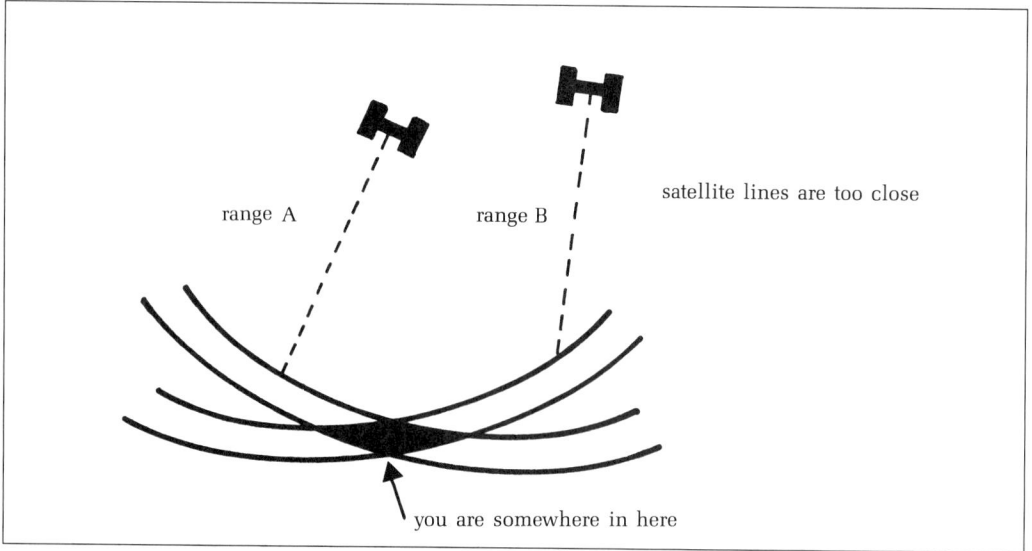

Poor satellite geometry.

geostationary over a particular place on earth; they have been placed in such precise orbits that they pass over the monitoring stations twice each day. At these times, the engineers can make allowances for any temporary changes in flight path, which might reduce the accuracy of the position fixers.

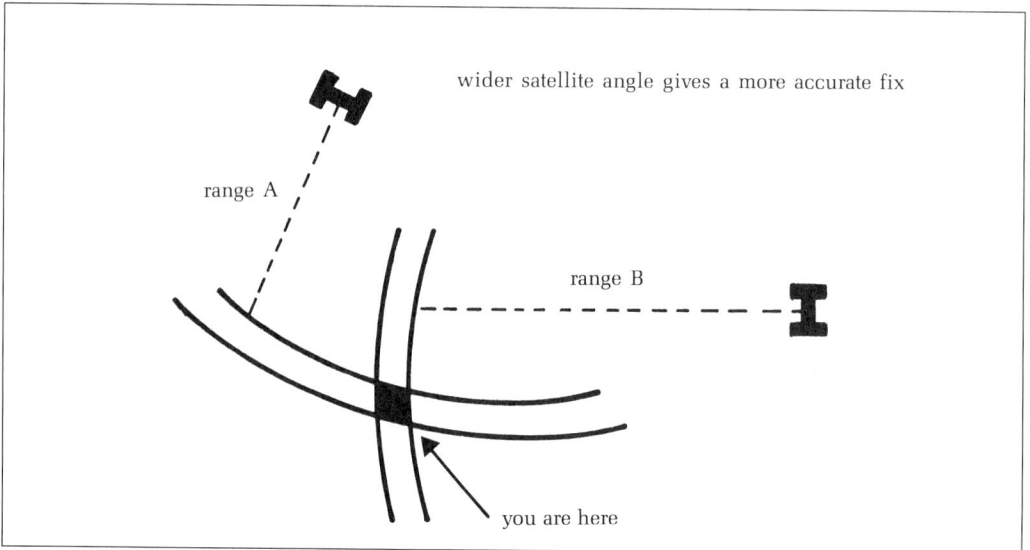

Different angles give a better fix pattern.

## The Atomic Clocks

Out in space, the most important equipment is the atomic clocks carried by the orbiters. This has nothing to do with atomic energy, but explains how the natural oscillating frequency of a particular atom is used as the clock's time-counting metronome. Atomic clocks are by far the most accurate timepieces yet devised. When they say that it is 12 noon, they mean just that and not half a second after noon, or even a millionth of a second before it.

Atomic clocks are also very expensive. Each one in flight has cost the US government about $100,000 but their absolute, zero-defect accuracy is a very essential element of how GPS works. Unfortunately, not all boats and planes can match the atomic clocks, so the scientists have devised some very clever methods to eliminate the deficiencies which might be caused by the less good quality time-keeping of even our excellent quartz-controlled clocks and watches.

## Basic Principles

The fundamental principles employed by

GPS can read to three decimal places or about 6 ft.

GPS are not new and are very well understood. Radio signals can carry a vast amount of information at the speed of light, which has been very accurately measured at 300 million km (186 thousand miles) per second. If we know the absolutely precise time that a particular part of the radio signal left the satellite and can record exactly when it arrived at the boat, some simple arithmetic (for a computer) will produce the boat to satellite distance. This distance might be anywhere along the arc of a circle whose centre is the satellite itself. It will, however, be a very precise distance because the quality of the information is good. As with Decca, if we add a second transmitter to create another arc, there will be two points at which they cut each other, so our boat will be in one of these two places.

Unfortunately, because the satellite is moving and because the clock on the boat might not be absolutely accurate, practical life is not as simple as theoretical trigonometry. If, for instance, the GPS receiver clock was only one-thousandth of a second out of synchronization with the satellite clock, our plotted position could be 186 miles adrift. The electronics engineers have proved equal to the task, however, and have devised a number of methods of ensuring accuracy to about 15m (50ft) all day and every day.

## GPS in Practice

Because we are dealing with clocks, let us think in terms of time and note our range from the satellite in the number of seconds it might take a signal to travel from satellite to aerial, rather than the number of kilometres between them. Let us look at an example of a simple position fix.

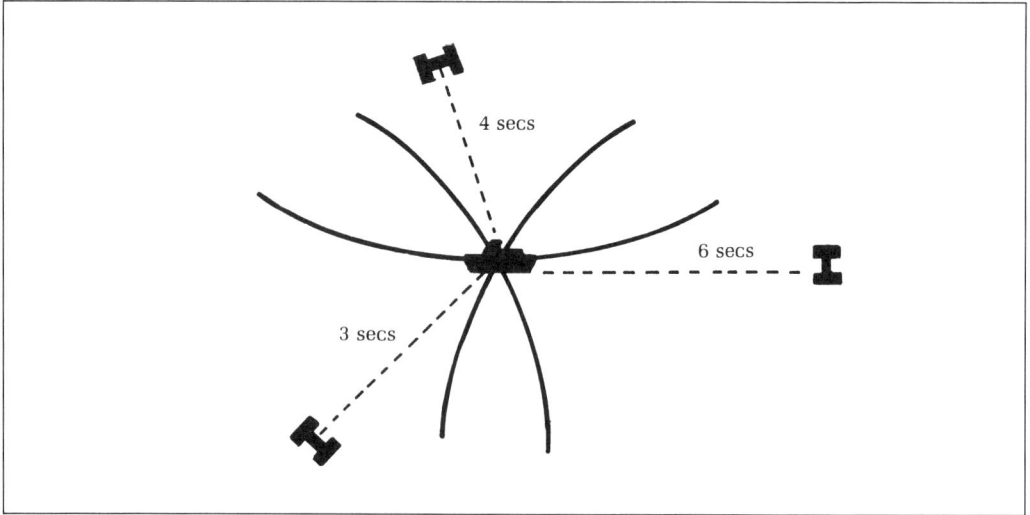

We measure our distance in seconds.

Suppose we are at 3 seconds from satellite A and 6 seconds from satellite B. The time/distance circles cut each other at point X, so that is where we are. However, if our GPS receiver clock on the boat is one second slow, the unit will think that it is 4 seconds from A and 7 seconds from B and will calculate our position as being Y. The skipper will have no means of knowing that the fix is way out.

The remedy is to bring in a third satellite signal which, in our example, we will suppose to be at a time distance of 9 seconds from the boat. Our slow computer clock will see this as 10 seconds. If we draw this circle we have a radiation diagram where the circles simply do not cut each other at all. The positive fix is the typical very large 'cocked hat'; the navigator's nightmare of a very large space. He knows the boat's position is in there somewhere, but where?

The GPS computer soon realizes that there is no such position as 4–7–10 seconds from the chosen satellites, so it automatically goes into a routine of simultaneously subtracting a microsecond of time from each of those received. The effect is to reduce the size of the cocked hat by drawing all its sides – and thousands of points along those sides – into its centre. Eventually the arcs will coincide very precisely. The computer then realizes that it has found a true position and quickly converts it into lat/long and all the other distance and bearing information we have come to take for granted.

This location is not as haphazard as my very simplistic explanation implies. Some positions might be nonsense because they are out in the stratosphere. The software only works on possibles and has some very sophisticated algebraic number crunching routines to make life easier for itself. Twenty years ago, all this would have seemed like pure science fiction in terms of speed, accuracy and the phenomenal amount of backup information and

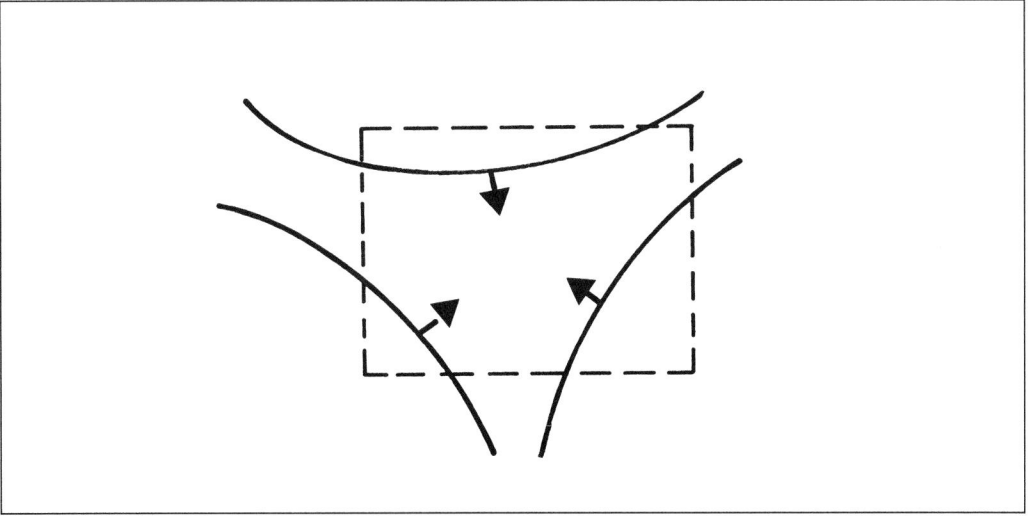

The computer chips away until it gets a logical fix. There will only be one logical place where all three lines coincide.

calculating power which can be stored in a module the size of a matchbox. With the present generation of microchips and processors, engineers might see it all as quite normal, but they still have to do some very clever things to overcome the problems and to design machines which non-engineers find simple to use.

## Pseudo Random Codes

One of the more clever devices is the so-called pseudo random codes (PRC), which allow the computer to work out exactly when the signal it is receiving actually left the satellite.

The PRC is not really a random mish-mash of changes of voltage pluses and minuses. It is more a complex cocktail of these things creating thousands of digitized pulses. Because it is complex (but calculated) there is no danger of ambiguity. The boat computer has access to a

'diary' which lists all the patterns. As soon as it recognizes a particular signal shape, it can make the comparisons and know at exactly which millisecond this part of the code was fired away from the satellite antenna. It can then calculate the distance refined to just a few metres.

## Satellites Will Not Stay Still

Obviously, there is one other important calculation factor which we have not yet mentioned. Because the satellite is moving, we need to know precisely where it was when it released the part of the pseudo random code which we are using.

Fortunately the orbits and the orbital speeds are very stable and have been calculated in advance. On those two occasions per day that the satellite passes over the monitoring stations, altitude, track and velocity can be measured to within a few metres. If it has been pulled off line

41

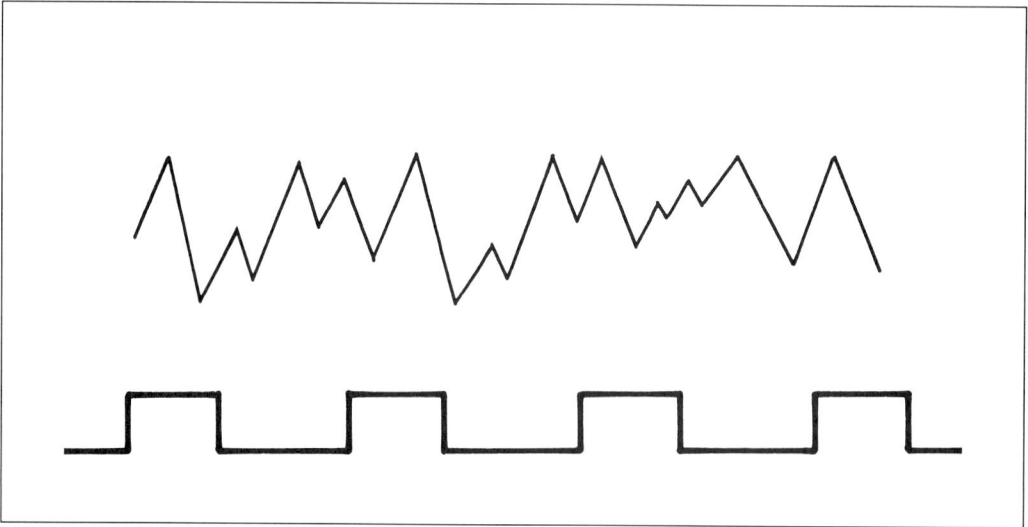

How the engineer sees a pseudo random code.

slightly by such interference as the moon's gravitational effect, the satellite is made to transmit this 'state of health and orbit' information as part of its routine data.

## The Ephemeris

In the GPS unit itself, the computer also has a catalogue of the satellite's predicted

| GPS | | | | | | | GPS status | | | |
|---|---|---|---|---|---|---|---|---|---|---|
| PRN | 0 | 1 | 2 | 3 | 4 | | PRN | AZ | EL | S/N |
| 0 | | | H | H | H | | 21 | 120 | 33 | 11 |
| 1 | | | H | | | | 6 | 40 | 18 | 47 |
| 2 | | | H | | | | 14 | 103 | 12 | 48 |
| 3 | H | H | | | | | 17 | 75 | 60 | 41 |

H = healthy usable satellite

At switch on, the GPS' receiver shows which satellites are available. Information on the right of the screen is where the satellite is located in relation to the boat.

A sequencing GPS receiver.

times and distances – the so-called ephemeris information which has long since been part of the navigator's armoury – and can interweave the current 'health' bulletin to calculate its positions. It also makes allowances for the known way in which radio signals slow down as they pass through the ionosphere.

There are other sources of potential error – changes in atmospheric pressure, raindrops, fog, thunder, terrain and even our clocks rounding off their calculations to the nearest millionth part of a second. Even with all these mini hazards and pollutants, you still have to admit that GPS is an incredible and advanced system. It is so good that if you were in an average high street, it might put you on the left pavement instead of the right, but mostly it will locate you in the middle of the road, or even better.

## Machines and Machines

With such potential precision, the navigator's ability to use this powerful information to best advantage must eventually be decided by the quality of the actual unit installed on the boat.

Unfortunately, not all machines are brilliant, so the seafarer needs to know exactly what he is buying. There can be

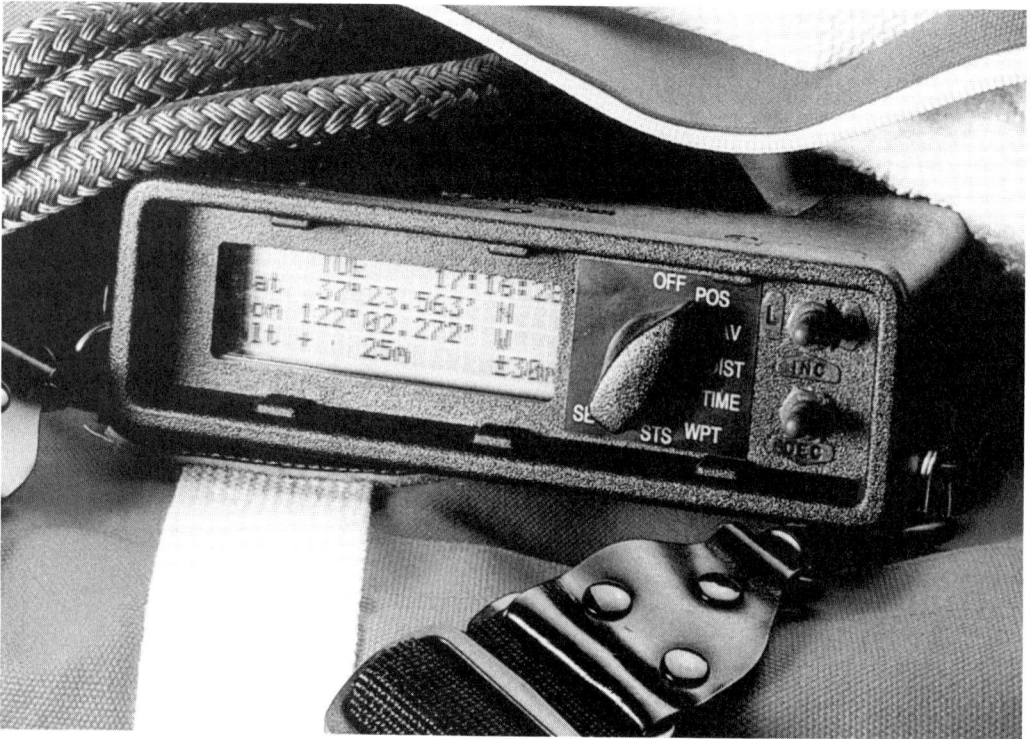

Fast multiplexers are excellent.

as much waste from being overgunned as being short on quality and having to upgrade.

**Sequencing GPS units** are the entry point. These are single channel receivers with only a limited amount of computing power. As we need information from at least three satellites to get a marine two-dimensional position fix, the single channel has to work very hard to switch from one to the next and to feed the collected information into that part of the black box which does all the collating number crunching. For marine use, we will do best to ignore those 'minimum power' portable units which only run off internal batteries. To save power they

frequently shut down, and so do not give a continuously updating fix. We need a receiver which stays on all the time.

**Fast multiplexers** are also single channel, but their circuitry is very sophisticated. They are less sensitive to clock slippage and fast enough to look at basic signals and data messages whilst simultaneously making ranging measurements in a fraction of a second. These are not cheap units because their sophisticated software is as expensive as adding extra channels.

**Two-channel sequencers** are the next step up. They have more internal signal acquisition power, so will lock on to satellites more quickly and are better than single channel receivers at using satellites which are close to the horizon. They will give better speed accuracy if you have a fast power boat. Inside a working two-channel sequencer, one channel can be sorting out the position information whilst the other is locking on to the next satellite to be processed. As in all GPS, more speed means more accuracy.

**Multi-channel receivers** can consist of anything up to twelve. They are simultaneously able to track all the satellites in view and to select only the best for strength and trigonometric position, whilst retaining enough spare capacity for extremely rapid calculation and total, instantaneous update even at aircraft speed.

Other features to look for are firstly to ensure that the receiver is programmed to align with the charts which you will use. The better models will switch between such chart datum references as the British OS36, the European ED derived from information gathered by the hydrographers of France, Spain and so on and WGS84 which is the 'language' of most United States charts. By getting into the right programme, you can transcribe straight from screen to paper without calculating those confusing offsets which are a feature of the user information overwritten on every Admiralty chart. These mostly advise you that the chart has been written to WGS84, so the advised position will need to be moved a certain

The multi-channel units are superb.

distance south and different distance east to be plotted on the chart. Happily, with present software progress this sort of work with parallel rulers and dividers is a thing of the past.

At some point you might wish to download waypoints into a video plotter, or a laptop computer, or even to upload them, or to have your lat/long displayed on your radar screen. This needs a special piece of software, included as standard on good GPS units. You can also let the GPS receiver control your autopilot, thereby putting the boat on totally integrated automatic control. This is a vast and contentious topic which will be looked at briefly in Chapter 11, but before we get there, let us learn to drive electronic navigators in their manual mode.

## SUMMARY

- The Global Positioning System (GPS) is here to stay, but you still need to learn how to use it.

- There are over twenty satellites orbiting at about 11,000 miles (18,000km) out in space.

- Satellite ranging is the proper term for the way the system works.

- You need three satellites in view to get a sea fix.

- The theoretical accuracy is 15m (50ft).

- Selective Availability (SA) can be switched on and off.

- SA reduces GPS accuracy to a 100m (330ft) radius circle.

- GPS is unaffected by season or weather.

- Multi-channel receivers are better performers, but are also more expensive.

- Check that your receiver is on the same datum as your chart.

# 4
# PRACTICAL
# NAVIGATION

Decca, Loran-C and the various forms of satellite navigation may have wide technical differences, but their similarities are such that if you can operate one, you will soon feel at home with another. Mostly, they share a common operational language, in which some very powerful functions are given different terms – largely decided by the side of the Atlantic which inspired their origin.

Navspeak is almost a shorthand language in its own right, but a couple of trips soon makes you fluent. You will continually be using some of the terms in the list below:

**BRG (Bearing)** The direction in degrees T to any chosen point whose co-ordinates are in the navigator's memory bank, or have been put into your route plan for the passage.

**CMG (Course Made Good)** The actual track which the boat has been following. It is a true course which can be drawn on the chart and which will often differ by a surprising amount from the compass

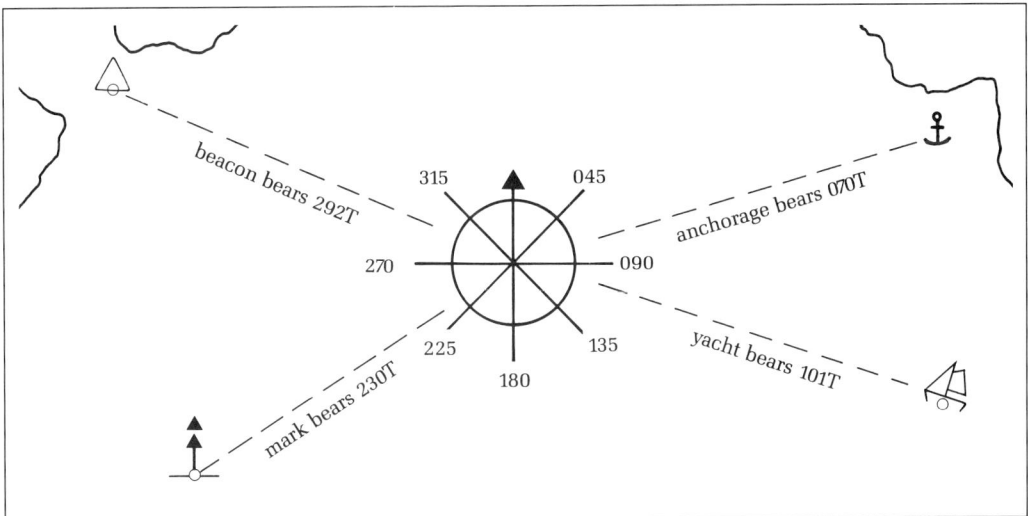

You will soon learn to use bearings.

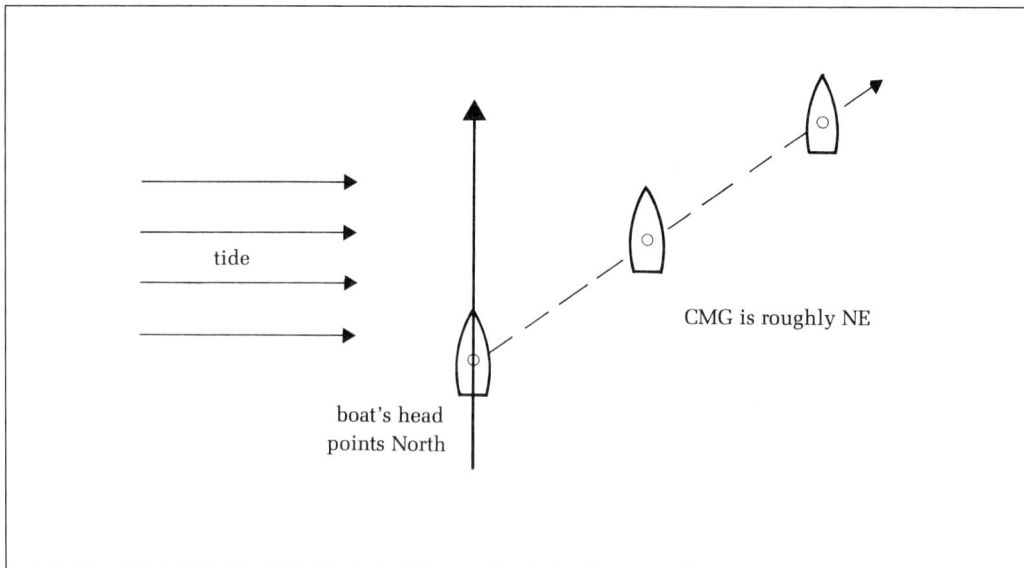

tide

CMG is roughly NE

boat's head
points North

CMG differs from compass course.

course. It shows how much the boat is being moved sideways by a lateral component of current, or how much leeway she is making downwind.

**COG (Course Over Ground)** Another way of saying CMG.

**CTE (Cross Track Error)** The direction and distance which your passage has moved away from the rhumb line, or straight line distance between any two points to be travelled.

**CTS (Course to Steer)** The direction indicated by the navigator in order to get to your next waypoint.

**DTK (Desired Track)** The ideal line to get to your destination. It approximates closely to bearing. Some navigation boxes take into account the boat's sideways crabbing progress and give the DTK as the optimum line to take to allow for this sideways drift. The moral here is to read the handbook carefully to see what it means for your particular model.

**ETA (Estimated Time of Arrival)** This is self-explanatory and will give an actual clock time for when you will arrive at your destination (next waypoint or final anchorage) according to how the machine is set up.

**ETE (Estimated Time *En Route*)** The time in hours and minutes (duration) of the current leg, or of the whole journey – according to set-up.

**GO TO (Go To)** A rapid way of asking the navigator to tell you the course and distance from your present position to any other place or point listed in the memory.

**HDG (Heading)** Closely akin to BRG and TRK, this is a direction to a given point.

**OCE (Off Course Error)** This is identical to CTE.

**RCP (Reciprocal)** A 180-degree reversal of any compass course or direction.

**ROUTE** The arrangement of a number of waypoints in the order in which you wish to travel around them. They may be listed

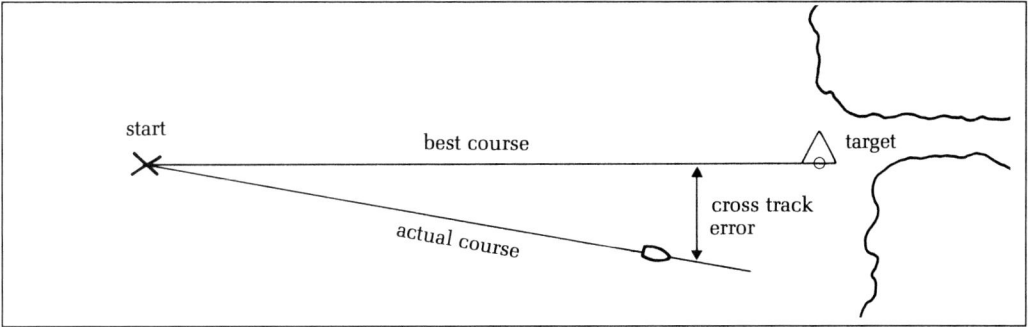

Cross Track Error. How far off line?

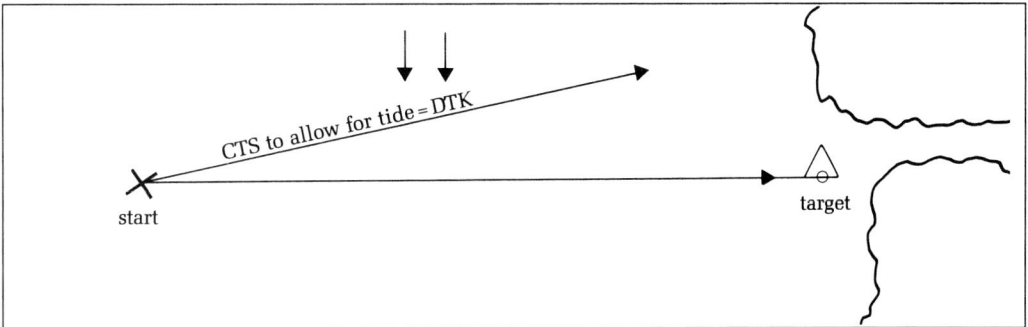

CTS and DTK are much the same and allow for drift.

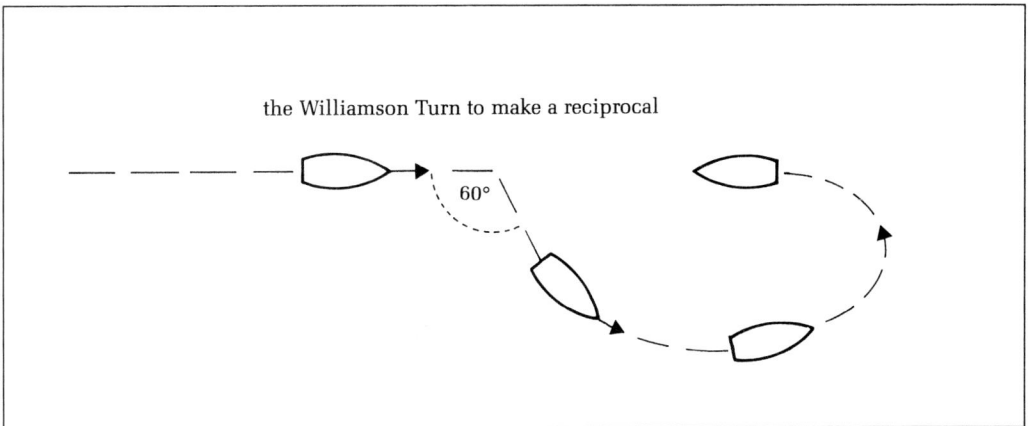

A reciprocal to rescue a man overboard.

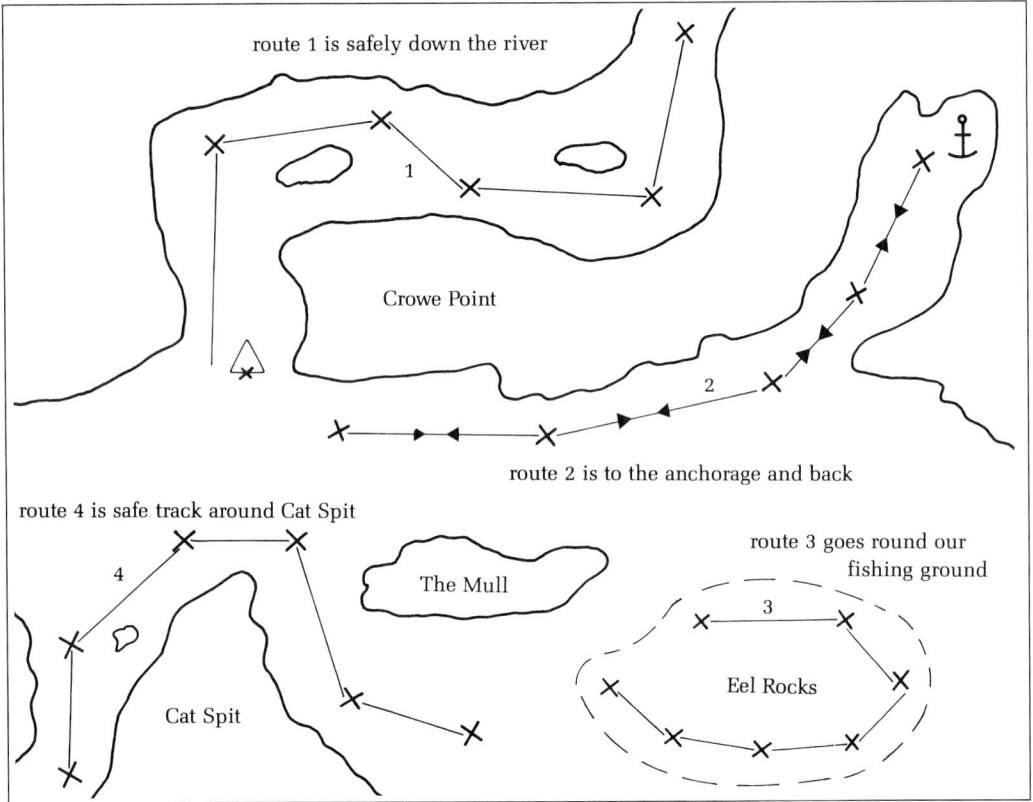

route 1 is safely down the river

1

Crowe Point

2

route 2 is to the anchorage and back

route 4 is safe track around Cat Spit

4

The Mull

route 3 goes round our fishing ground

3

Cat Spit

Eel Rocks

Routes can be stored in the memory.

in a different order in the navigator's waypoint bank. Some models can store routes as a separate module, which can then be inserted *en bloc* into a sailplan. A channel to be negotiated half-way through a passage is a good example of a stored route, while the usual run up to your own harbour is another.

**SAILPLAN** An arrangement of way-points to cover the passage which you are making.

**SPD (Speed)** This is either the boat's rate of progress as shown by the paddle-wheel log, or the actual speed over the ground shown by the navigator. Some models can

The paddle-wheel log computes speed through the water.

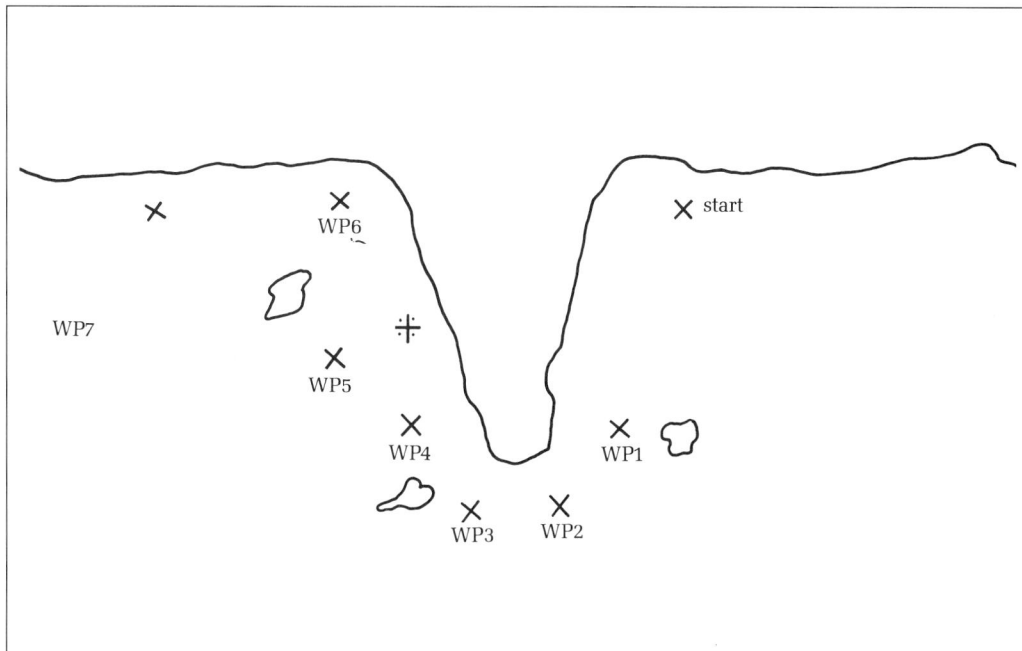

A series of waypoints take you around a headland. To make the passage safely we shall need 7 lat/long co-ordinates.

convert this from knots to kilometres per hour, feet per hour, or even to metres per second.

**TARGET** This is another name for a waypoint. In radar language it is any echo which appears on the screen.

**TRK (Track)** This is similar to CMG or COG.

**UTC (Universal Time or GMT)** This is the actual time used by navigation systems. It ignores local time differences such as daylight saving time or the hour difference which normally exists between Britain and Continental Europe. It is referred to as Local Time, or Clock Time.

**WP (Waypoint)** Any point along your route. It is also used to mean any place or position which is stored in the navigator's memory.

**XTE (Cross Track Error)** This is identical to CTE or OCE.

**ZONE TIME** This can also be called 'clock time'. It is the actual time applicable to the present season and location. It adds on any daylight saving time, or can take into account the differences between one country and another.

You will also rapidly get used to the multiple functions and the immense power which your computer gives you. However, basically all electronic navigators do the same thing:

1. They tell you where you are at the exact present time.

2. They give you enough data to plan a passage and to calculate the time *en route*.

3.    They tell you exactly where the ship is going and when it will get there.

4.    They keep a record of where you have been.

In order to see how this works in practice, let us plan and make a theoretical journey of, say, twelve hours from my home port down a fairway, across some open water, then go via a tidal channel and into a strange harbour. It is the route shown on the accompanying plan (page 57) and is almost a condensed version of the passage from any British south coast port around the corner of Brittany, inside Ushant and using a French harbour as a launch pad for a trip across Biscay.

To make life more practical, let us do this trip using the Decca and the GPS which happen to be on my boat as I write, but do note that the method would translate to any other type of machinery. Typically, let us program in to the computer that we can do 5 knots under sail and 8 knots motor sailing in still water.

# First Principles

All my own trip planning is based on a couple of precepts forced on me the hard way – from the experience of making silly mistakes which caused that panic of uncertainty, and which froze my brain when I did not know exactly where I was.

First, it is much easier to work out positions, to read charts and to write up passage plans on a still surface, rather than when the boat is banging about and you need one hand to hold yourself steady and at least one eye on where the boat is going. I therefore do mine at home, or on board whilst I am still in harbour.

We all eventually come to rely very heavily on our electronic navigators, and this is how it should be as they are very safe. However, these same safety factors can bring dangers; the black boxes are only as good as the initial data which they are given. Remember the 'garbage in: garbage out' warning, so always try to do all your planning very thoroughly and cross-check it before departure. Also remember that the boat's skipper can never be overinformed. You might not expect to be close to a particular lighthouse or an adjacent *en route* rock, but things do not always go according to plan. It is then a great help to have all the information immediately available. So do it now, whilst you have plenty of time and an otherwise uncommitted mind.

# Lets Go Cruising

## *The Cruise Plan*

Drawing this on the chart is the first practical step. This lets you see that you are not passing over any shallows, or other submerged hazards and that you have left enough margin for error. The tide might not do exactly what the almanac predicted for the day, or you might have a small on-board problem, so you will be glad that you have been a good enough skipper to leave yourself a bit of sea room, just in case.

There are some skinflints or 'house proud' skippers who suck their teeth in horror when I advocate drawing on their expensive charts. Professional navigators always mark up the chart so that everybody on board is aware of what is happening and so that good information is to hand if a panic occurs. I draw in course lines and add the course and distance

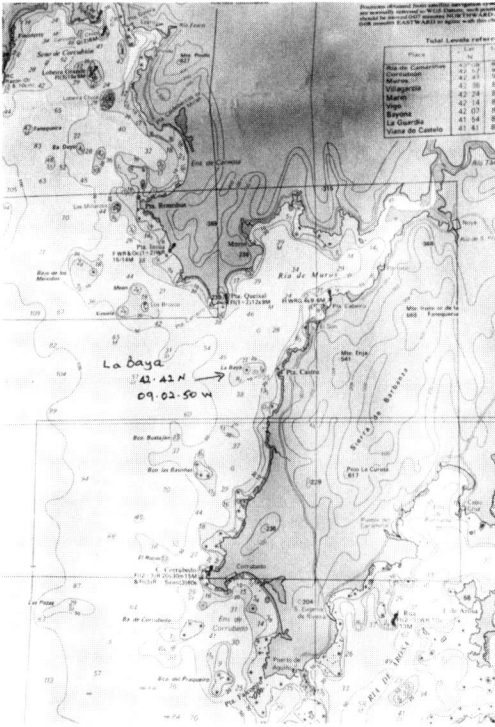

Note the latitude/longitude of hazards in the log and on the chart.

of the two. My best navigation screen will accept two lines of twenty letters, so a full name or a short description can be entered, for example, 0.75 miles off Anvil Point. This information is typed to the screen from the keypad.

Both nomenclature methods can be found on any number of makes of electronic navigators and neither can really be said to be universally better than the other. Simple numbers are easier to key in, but the synoptic names are a fast *aide-mémoire* when you are trying to do six things at once.

The keen navigator will always keep a separate Waypoint Catalogue of every position which he has taken off the chart. Electrical faults have been known to wipe out the electronic memory (even though this is rare) and you might need the waypoints on the return journey, or even

We also keep a waypoint book.

between each waypoint and even note the lat/long of some *en route* hazards and other points of possible interest. This same data is also on a piece of log paper and is entered into the log as I go.

## The Waypoints

These can now be worked out as lat/long and given names and sequential numbers for the route, or for storing in your navigator's Waypoint Catalogue. My two on-board systems use differing storage terminology. Some units simply designate each waypoint by a number, whereas the others give each a short name composed of five letters, or numbers, or a combination next year. Anyway, it is also fun to look them up again when you are reliving your cruise on a dark wet January evening at home.

## The Draft Plan

This is a set of rough notes on the course and distances you have worked out between the waypoints, together with a few references of how far you should be off any useful (or hazardous) landmarks. It is a task for pencil, chart, paper and dividers, plastic plotter and a clear mind.

## Entering the Data

Putting the data into the navigator can be done at home with a 12V power supply, or by using the pack which many manu-facturers include as a combined mains and recharging unit. ( A number of companies manufacture a portable, long-life battery pack, which is useful for all sorts of on-board and transfer tasks.) Once all the information has been entered, it can be juggled into a route (sometimes called a sailplan) using a combination of the new references and others which might already be in memory. You are now ready to check the electronic efficiency against your own navigational skills.

## The Double-Check

This is where you scroll the navigator through every leg of your proposed voyage and verify that it approximately matches your own course and distance for each of them, and is probably the most important

One GPS has its own battery, so we program it at home.

planning and safety factor of the whole operation. It is at this point that any incorporated errors will show up and it is vital that we see them now rather than when we are out in tide or fog. These errors are all human and we all make them.

Commonly, somebody misinterprets a lat/long reference by reading from the wrong side of a chart line, or by making an arithmetic error when counting off fractions of a mile at the chart edge, or even misreading the scale. It is also easy to misalign a Breton Plotter, or to twist parallel rules if you use them.

Keyboard errors include pushing a wrong letter or number, or even mishitting a key so that the digit does not record, or not watching the screen cursor and putting in wrong information. It also happens that north and south, or east and west, are not correctly set so you are aiming at the wrong global hemisphere. It is only by checking your manual navigation against the electronic that you will be totally sure that the courses you will be using are correct and safe.

It is worth repeating that it is a good plan to work everything in degrees True (T on the chart), which is the same as the notation used by the electronic machinery, and to let the helmsman add the magnetic variation. As you write on the chart, it is also a good habit to denote nautical miles with a capital M, which is the same symbol used for the number of miles in a

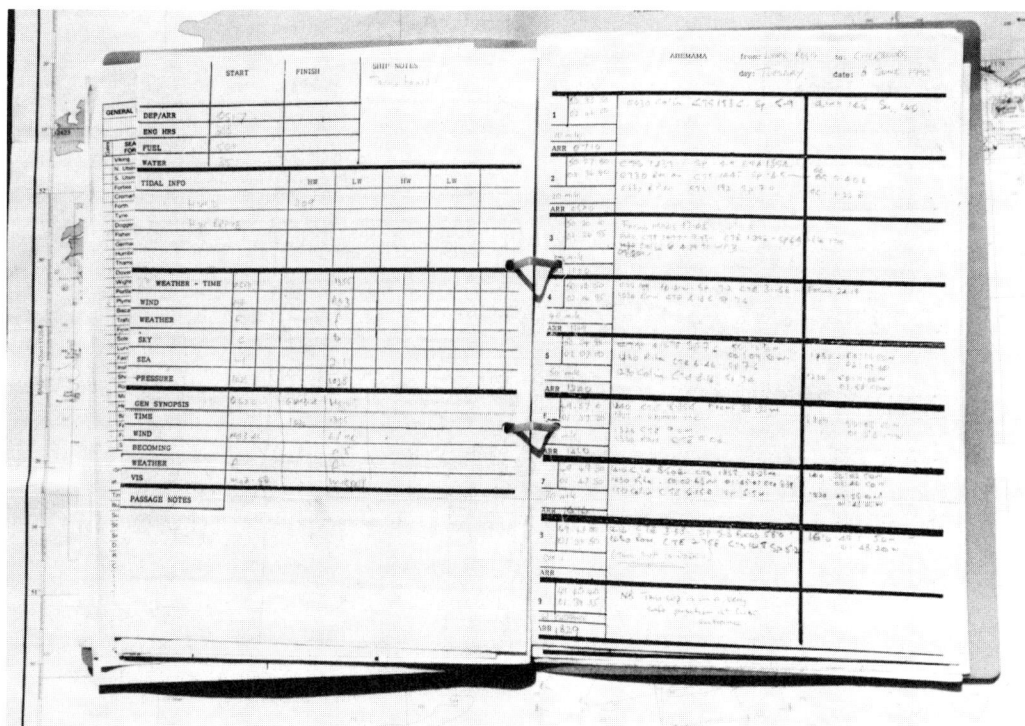

A completed waypoint plan sheet.

lighthouse range and distinguishes miles from metres which are normally symbolized by the lower case letter 'm'.

Radio navigators and their satellite equivalents are marvellous pieces of equipment, but only a genius or a fool does not check and double-check how well he is using them.

## The Final Plan

On my own boat the final plan consists of a good copy of all the waypoints for the journey, and the distances and bearings between them. From it, I can see the next course to steer to waypoint two even before the navigator puts it up on the screen as it passes waypoint one. If the course change is a big one, I usually start to make it gradually as I near the turn point. That eases the strain on the boat, the autopilot and the tiller arm.

On the photocopied planning sheet, I make a note of any lighthouses and other useful bearings along the way. If I am going into a long and complicated harbour or channel, I also have a brief note of what the navigation marks are – tower, buoy and so on – any distinctive colours and other interesting features; if coming in at night, lights or special marks; plus whether I need to leave it to port or starboard and by what approximate distance. As I pass each one in turn, I tick it off and note the time. Then I never have the panic of wondering whether I have passed Brent Tower, or if it is the one just coming into view, so that I am about to run aground. This is especially important if you are running into a relatively long, multi-buoyed harbour like Poole, Southampton or Saint-Malo. Later, the same information can be used to plan and to time my exit, or it can be used for the return journey.

## Come the Day

Many electronic instruments are very susceptible to voltage 'spikes' – those dips and peaks of electrical tension which can be caused by such things as switching on diesel heater plugs and starting a big engine. Outboard motor alternators are especially prone to creating such surges and to causing interference at start-up and just after. On many boats, it is good practice to have all the electronics shut down until the engine has started and the charging rate has settled. This only takes a couple of minutes, then the navigators can be left to lock on to their information systems whilst you do all the other jobs necessary to get the boat ready for sea.

Some electronic position finders need to be given an approximate position to make their transmitter station search easier when they are switched on for the first time. Others have an auto-locate function which will find itself from zero input, but this is rarely as good as the manufacturer's brochure claims for shore-based radio systems. One of my own Decca machines rarely betters 40 minutes to auto-locate, even when the nearest stations are giving high signal strength readings. Because of its nature, GPS is much faster to get itself sorted out. Three minutes or less is a typical time from switching on to getting the green light for navigation.

Once the traffic lights (or other indicators) show that all is well, I usually flick through a number of functions to check on things I like to know:

1. Signal strength and signal-to-noise ratio.
2. That I am on the correct chain and what is the current circle of uncertainty.
3. The battery voltage level.

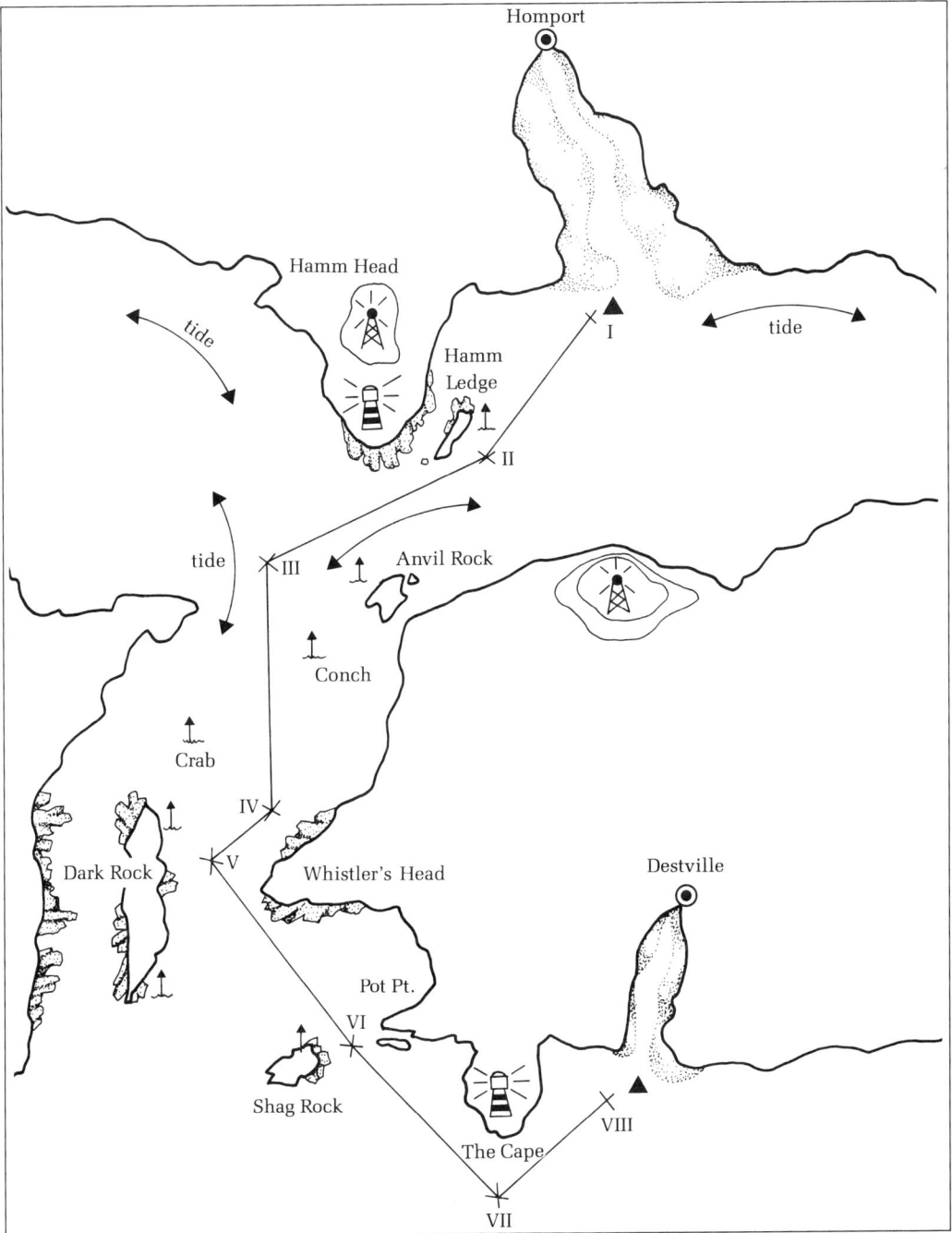

The route for our cruise.

4. That the clock and date are correct.
5. That I am on the correct chart datum.
6. That all resettable distance logs have been set to zero.
7. That the anchor alarm is off.
8. That the desired alarms are on – antenna fault, voltage problems, poor signal strength, waypoint arrival warning and so on.
9. That the navigator is showing the course and distance to the first waypoint.

## Being Under Way

Clear of the harbour, I always run through the most often used functions – course over the ground, speed over the ground, distance to the first waypoint and running time to get there.

During this present 'paper cruise' I have made an imaginary 'own' harbour's Fairway Buoy the first waypoint. I know it well and it is reasonably close, so I should be bang on track when I reach it and can just check that the navigation boxes agree. I note that Decca and GPS are within 100m (110yd) of each other so all is well and I settle the boat on 190T for the next 15 miles to waypoint two, which I have marked about 2.5 miles clear of Hamm Ledge.

As I get into this leg, it is apparent that the tide is behaving as predicted and flowing north and east at 1.5 knots into the bay at this time. Both Decca and GPS are indicating that I am only making 5 knots over the ground even though I am motor sailing. This speed is less than the boat's log, which is hovering around 6 knots because its paddle-wheel is being turned too fast by the current passing under the keel. A 1 knot component of tide is slowing us and 0.5 knots is pushing us out to the east.

## Being Off Course

This straying away from the intended line shows up on the electronics. The Decca (it could be Loran-C) has a hook symbol which indicates the direction to steer to get back on to the rhumb line and how far off track I have strayed. This is a clear, common display convention.

Some GPS units have an arrowhead pointing up and this moves sideways across vertical bars to the side you have strayed. The space bars can be set to varying distances from 200m (220yds) to 1 mile. This means that you have to estimate the actual distance, but it is simple to do and I personally like the way the arrowhead reverses to show when I have passed a single waypoint (say a fishing mark or dive site) and that it is now showing the bearing and distance back to it. More expensive navigators incorporate a simple track plotter. This displays the waypoints as flags and follows the boat's course around them. If, for instance, you have set your CTE limits to 1 mile, two parallel lines will appear on the plotter, aligned with your intended course. By using an in-built zoom feature, you can get a good picture of where you are and where you are drifting in relation to the next waypoint.

In the old days, one should probably have drawn a tidal vector, which would angle away from the rhumb line and give us a heading which would finally converge with the destination. On a long trip, which might take in three or four tidal periods, this would be a very inefficient way to drive a sail boat or a motor sailer.

Because the boat is crabbing sideways my Course Made Good is showing 185T even though the ship's head is steady on 190, so the bearing to the waypoint will be

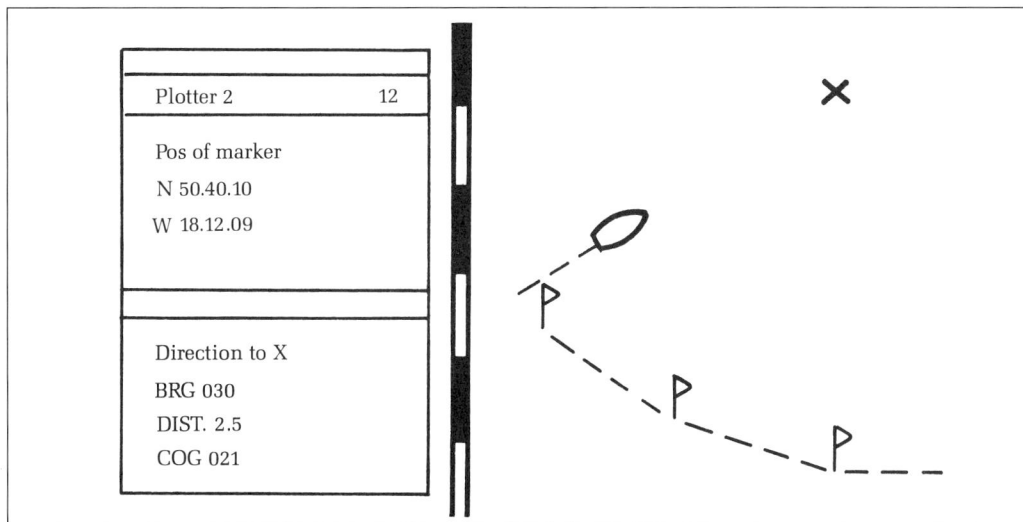

```
Plotter 2          12

Pos of marker
N 50.40.10
W 18.12.09


Direction to X
BRG 030
DIST. 2.5
COG 021
```

A typical GPS plotter screen.

gradually opening out. This is nothing to worry about as the distance to it and the ETA are both counting down at a nice, steady rhythm.

## Getting Well Off Track

I am not too concerned about being off my rhumb line to the east. First, that gives me a bit more margin of safety clear of Hamm Ledge and I shall soon pick up the ebb tide and will see the speed creep up and the Cross Track Error come down as I am pushed west again.

Out here on the open sea, it would be a mistake to try to hug the theoretical course line. To do so I should have to turn more into the tide and would consequently lose at least another 0.5 knots of speed. Then I would get over the other side of the track and have to fight the tide in the other direction.

Many fast boat drivers become paranoid about gluing the keel to the track line.

They are forever tweaking the wheel and the autopilot a few degrees, but by the time they have reacted to the navigator's indications, they have often crossed the track line again and are steering a very unseamanlike, zigzag course.

I shall be content to let the boat have her head, to plot my position on the hour, every hour and to mark it up on the chart and write all the data about speed, Cross Track Error, Course Made Good and so on in the log, so that I can give the electronics good data again if a fuse is blown and they shut down. I could also take over manual navigation from that known position and time.

I have set one of the navigators to sound a reminder two minutes before each log time and it also warbles a warning that a weather forecast is imminent.

This leg is being sailed entirely on the electronics, but it is also good fun and good practice to plot the position by dead reckoning and by hand compass sights on

Hamm Lighthouse and radio mast and back to the Fairway Buoy. I can also get distance off the coast by radar reference. It is a good feeling when you can say to yourself 'It all checks'.

## From Two to Three

From Hamm Ledge to waypoint three my Desired Track (DTK) as shown by the GPS is 220T, but I am now being pushed south of my line. I am expecting a tide change along the way, but nobody can be precise about this to the nearest twenty minutes. I also need to arrive at the waypoint at no more than an hour after high water slack tide to get my ebb through Seasnake. However, because this is an electronics boat, I have plenty of assistance available.

## Using Two Navigators

On one of the navigators (say Decca) I can put in the lat/long of the Anvil North cardinal mark, then 'draw' a 0.5 mile exclusion circle around it. The machine will then sound a warning if we trespass inside this guard zone. The other machine can be left permanently with its screen showing Course Made Good and speed over the ground.

I can also get one to flick periodically to ETA at the waypoint and the other to show the Estimated Time *En Route* (ETE). By keeping tabs on my progress, I know whether to let the boat potter on even if I have to hang about for the ebb, or whether I will need to drive her a bit harder to get there in order to catch the tide. With so much assistance, all crew members' eyes are freed to watch for lobster pots and ships coming up around the corner on the flood. We are also freer to do the chart plot and to brew the tea.

## Down Through Seasnake

An electronically assisted ride down through a 6 knot tide channel with two of you helping with the pilotage is a mixture of quiet excitement and great satisfaction. With one set of eyes on the boat and watching for marks and the second brain occupied with the navigators, chart and the marker buoys, it represents much of what is best about cruising.

As we hit the top of the channel, put the boat on 180, and trim sails and revs for the most comfortable speed, the on-board dialogue might well follow the pattern below:

'What's the first mark?'
'Conch. West cardinal. Yellow Black Yellow. Two miles fine on the port bow.'
'Got it. Running time?'
'About thirty minutes according to the GPS. Then look for Crab – an east cardinal on a steel tower. Black Yellow Black.'
'I can see that one too. How far beyond is waypoint four?'
'One mile. Then you go on to 225T. For half a mile.'
'In this tide can I pick up 225 early?'
'Start to come round about 0.25 miles from WP Four. I'll put a "Go To" on Whistler's Head as a double-check. Come 225 at 0.75 miles from it. That should put you just to the north of the first of the two marks on this side of Dark Rocks.'
'Is that the White Tower?'
'Yes. We go about half a mile east of it.'
'I think I can see that one too. All the gauges OK?'

## Round the Corner

By this sort of conversation, one crew member is keeping the other on his or her

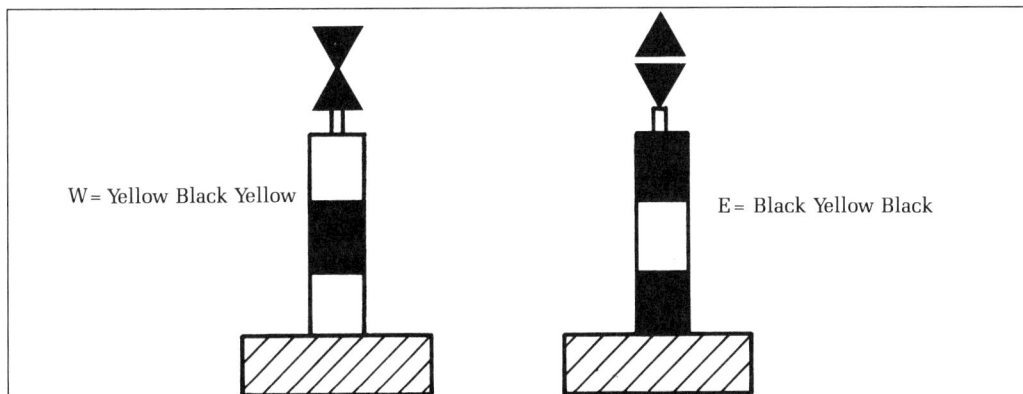

W = Yellow Black Yellow

E = Black Yellow Black

East and west cardinal marks are easy to identify.

toes and making sure that neither of us gets so mesmerized by Decca, GPS and the other gadgetry that we forget to look out of the wheel-house window, or to run periodic checks over the engine gauges, anemometer, echo-sounder and the sail trim.

In passing through a tight spot, where we need to hit the very centre of the channel between Pot Point and Shag Rock, the electronics and eyeball combination really come into their own. The first gets you on line and to verify your skipper's judgement, and the second keeps you on line and the bows bang on track.

Such a passage is not really typical because I have crammed a lot of navigation into a small period of time and distance to illustrate how modern electronics can make life at sea more interesting. Ten years ago, a cruising yacht would probably not attempt such a passage in poor visibility. Nowadays, with big-ship navigation equipment down in size and price to our level, we do not go looking for fog and rain, but if we get caught out in it, we can expect to cope – in other words, we are safer.

## Almost There

As soon as I can let my concentration diminish a little, I shall begin to have another look at the channels up to Destville and to think about how and where I shall berth. After arrival and when the boat is snug to her anchor, I then go through the closing down routine, which is almost as important as the planning.

The first steps are to put all the mechanical notes in the log – engine hours, fuel state, barometric pressure and battery voltage. Next I note the direction of the wind and tide and check them against the boat's head as shown by the compass. A note of time and water depth will also be useful. Then I can pick out some shore marks to give a quick visual indication should the boat begin to drag anchor. This physical safety picture is backed up by noting a few headings and distances to piers, rocks and so on as indicated by the radar and is rounded off by a note of the precise lat/long of the anchorage. This not only for use on another day, but also as an increased safety precaution, which could be further 'tightened up' by setting a

Decca/GPS alarm to sound if the boat drags or swings more than 100m (110yd) from the spot, and by echo-sounder alarms to warn me if the water becomes either very shallow or very deep.

These precautions are not being 'over-fussy'. Firstly, most serious navigators enjoy such mechanical discipline and, secondly, I have often been glad to have such good information available when I have had to lift the anchor and move to a more sheltered spot in the dead of a dark night. As soon as the anchor breaks out, I immediately have a safe course to steer and can avoid any hazards, even though I cannot see them. I do not always set all these on my own boat, but prefer to reset all the distance logs to zero again and to shut down all unnecessary electrics for the duration of the stay.

The final electronics act is to turn the battery bank selector switch on to the setting for the huge bank of domestic batteries and to isolate the other so that it

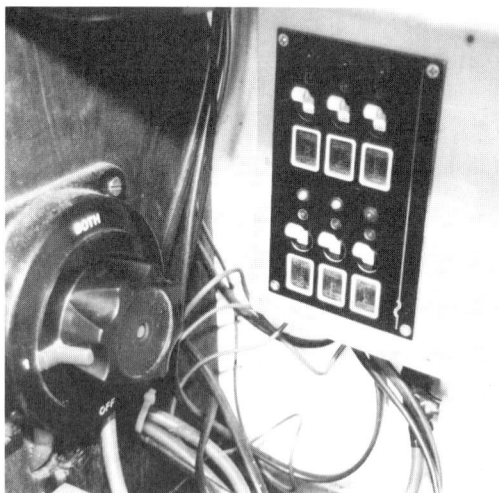

The final act is to set the battery switch to the domestics.

is not being discharged by the refrigerator, lights and so on so that it is always available, fully charged, to start the engine.

---

**SUMMARY**

- All navspeak uses a common set of abbreviations.

- Passages must be planned on paper, then put in to the electronics.

- It is easier and safer if your planning is done at home or in harbour.

- Do not be afraid to write on your chart; that is partly what it is for.

- You cannot plan a trip without reference to a tidal almanac.

- Always have somebody to check your entered data.

- Make yourself some planning and log sheets.

- To be very safe, shut down the electronics when you are engine starting from cold.

- Talk the journey through with your crew so that they can take over in emergencies.

- Electronics are marvellous, but you must still look out of the window.

# 5

# THE ELECTRONIC COMPASS

An electronic compass is a very welcome addition to the equipment on my own boats simply because it makes many things so much easier to do. However, even though I am 'electronics orientated', I have not yet reached a point where I would consider boat electrics to be so fail safe that I would even contemplate going to sea without having on board a magnetic compass which is correctly calibrated and swung to the particular boat upon which it will be used. Indeed, I use its tried and trusted reliability for most of my navigation, but let the electronic heading sensor do those jobs for which it is eminently more suitable.

The amount of specialist work which I ask it to do increases with time and with the new engineering systems being built into powered compasses, spiced up with the new techniques which I learn continually. Also, because digital and pulse information are easily generated and transmitted by the electric compass's own systems, it is ideal for such applications as integrating with a radio navigator, putting directional information up on the radar screen, or sending it to repeaters on the flybridge and down to the below decks nav station. Next to a fully stabilized gyro, the electronic compass is unsurpassed for keeping an autopilot on track.

## What Can It Really Do?

To know which tasks it can best undertake, you need to understand how the electronic compass works. It is also sometimes called a fluxgate compass, which gives away the trade secret that it is driven by a flux element, a coil, or a toroid. In simple terms they mean much the same thing.

The flux winding element is mounted on a pivot, or suspended on a plumb bob and is excited by a small electrical charge. This in turn creates a back voltage, which is at its maximum when the unit is aligned north–south along the earth's lines of magnetic force. As the boat and compass turn east, this voltage diminishes progressively in absolute inverse proportion to the size of the angle with north – in other words, the greater the angle, the less the voltage.

Because the voltage mirrors the amount of deflection from north exactly, the unit is able to sense precisely where it is in relation to that pole. North gives maximum voltage (Vmax), east gives a minimum signal (Vmin), and the signal rises again to Vmax as the turn continues to reach south. In one complete 360-degree turn, there are obviously two Vmax and two Vmin points, which the engineers untangle by

Inside a fluxgate: complex but reliable.

the electronics software, or by having two flux units tracking each other.

This system gives the fluxgate compass a number of advantages over the magnetic card mounted in its bowl of liquid.

## The Advantages of Speed

The first advantage of the fluxgate compass is speed. It comes as a surprise to most of us to learn that the adopted industrial standard for the time a good compass takes to settle, following a rapid 360-degree rotation, is in excess of fifty seconds. Every mariner has experienced the frustration of waiting whilst a hand-held compass comes around to the bearing he plans to take, then oscillates on either side of it. Next, just as he thinks that he has it settled, the boat lurches and the swinging starts again. The same phenomenon occurs when you are trying to settle the boat on a heading in waves.

With a fluxgate direction finder, however, this does not happen. You can spin it like a top and it will still settle in a fraction of a second. Whilst you swing the boat on to a reciprocal, or on to a new tack, the electronics track the boat's head and will give a very accurate direction just as soon as the turn stops. This makes the fluxgate compass ideal for use on fast boats, especially in waves. Honest skippers will admit that it is virtually impossible to keep a power boat within 5 degrees (or even 10 degrees on some occasions) of the intended course, especially if any sea is running.

My own dive 'taxi' is a 7m (23ft) rigid hull inflatable, which can be driven very accurately even in big waves when it is run on the electronic compass. The compass

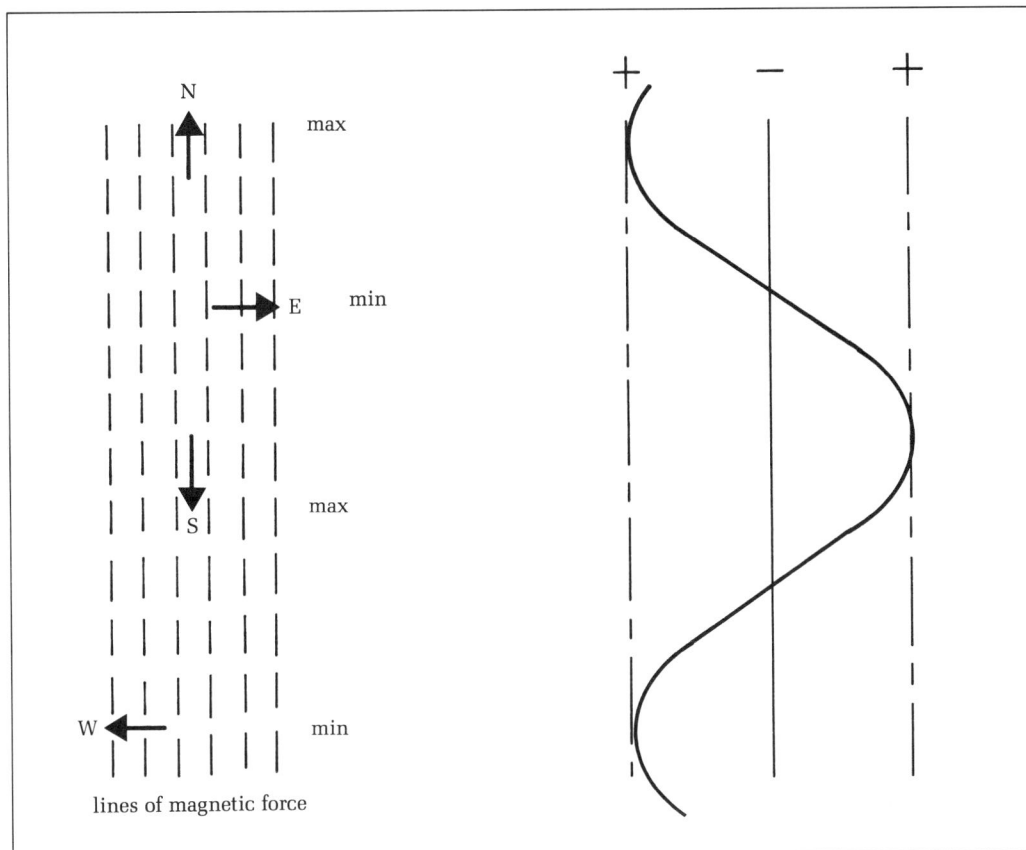

N and S give Vmax. E and W give Vmin.

has also taken this boat across 125 miles of open water and put me bang on to my targeted French lighthouse without the aid of Decca or Satnav. This usage alone would make it a good buy for any boat.

## Split-Level Navigation

Another set of advantages is created by the fact that most fluxgate compasses are in two parts – the actual electronic black box is separate from the display. This means that you can install the more delicate module down in the bowels of the boat, where it is protected from the elements and can be sited close to the boat's centre of pitch and roll. Such a location, with its lessened motion, would be best for any sort of compass, but for the fluxgate it is especially important. One of its weaknesses is that the sensing coil must be kept parallel to the earth's lines of magnetic force to maintain directional accuracy. This is why it is either mounted on a pivot or suspended in space. So, the more you can protect it from excessive movement, the better it will perform.

Newcomers to electronic navigation in

The fluxgate compass is in two parts.

power boats often think that they have a faulty model because the readings go haywire when the boat accelerates hard. They fear that the alternator is surging, or that they have an electrical fault. In fact all that happens is that the coil trails on its suspension as the boat picks up speed rapidly. Once the boat's velocity has settled, the readings return immediately to their normal and phenomenally rapid accuracy.

This limitation, imposed by the need to keep the coil horizontal, also occurs when using hand-bearing fluxgate compasses. It is disturbing to note how a few degrees lateral, or front to back tilt causes a major change in the read-out. Many people never master the skill of holding an electronic hand compass flat when the boat is rolling. You must balance it a bit like an experienced toper never spills his wine even when the glass is full and the boat is rolling. An automatic sense keeps the glass level. It can work with the compass too – and it must be made to work if your sightings are to be of any real use.

The two-part configuration also makes life easier when you are installing the electronic compass at the steering position. Because the display unit contains no magnetic elements, it can be placed close to radio loudspeakers, radar units and electronic navigation systems without being thrown out of alignment. The sensing unit can also be made to drive any number of displays. This lack of magnetism at the read-out makes the fluxgate compass ideal for twin-engined power boats. It is common practice to mount the compass in the centre of the console, amongst all the dials and other electrical parts. At installation, a professional compass adjuster sets the compass with compensating magnets and card realignments. If the boat develops a fault and one engine has to be shut down, the magnetic compass becomes useless – and at a time when you might need it most.

Good quality fluxgates rarely need specialists to help with installation and nor do you need to go through the rigmarole of anchoring the boat steady on known lines to 'swing compass' and to rectify any inaccuracies caused by other equipment in the boat.

## Automatic Deviation Adjustment

My own electronic compass – like many – has an automatic deviation program built into its software. As the boat is driven through a full 360-degree circle, the unit senses the amounts of lateral and of fore and aft deviation and the 'computer' automatically rectifies the compass readings to compensate for them. This same function can also be used to put in local magnetic variation for navigators who do not like doing elementary mental arithmetic. If your boat always stays in the same area, this is probably well worth doing.

The skipper can also select from a wide range of display types when he is purchasing the fluxgate model: swinging card, digital, parallel graticule, or straight digital. Some compasses even permit you to change the display style as you go

along. This can also be an essential part of using its many functions. Like most electronic navigation aids, the fluxgate compass is a very versatile tool.

## How Many Functions?

I have top of the range compasses fitted both to my dive boat which will do 40 knots, and to my cruising motor sailer whose maximum hull speed is about 8 knots. On both vessels the multi-function screens are very useful and are typical of most good units. At power up, the unit automatically performs a self-test. If it locates a problem it will usually show its nature as a screen message. Once it has settled, a single push of a key enables the first function mode.

**Compass mode** shows the direction of the boat's head as an LCD read-out in standard three-figure digital notation. A second touch augments this with an analogue display showing heading in relation to north, which is itself one of the fixed items of the display arranged around the screen. The single spike can be increased to show all the degrees from north to south.

If I am steering due east, this will be shown as 090 in digits and as a screen edge marker in the 3 o'clock position. Alternatively, the entire 12 o'clock to 3 o'clock right angle segment of the screen edge display will show all 90 degrees as LCD spikes.

**Off-course mode** is entered by a single key stroke. It is probably where I spend most of my running time. The digital read-out in the centre of the screen can be set to show the direction of the boat's head. At the top of the screen, analogue spikes light up on the appropriate side of the 12 o'clock position and spread out to show

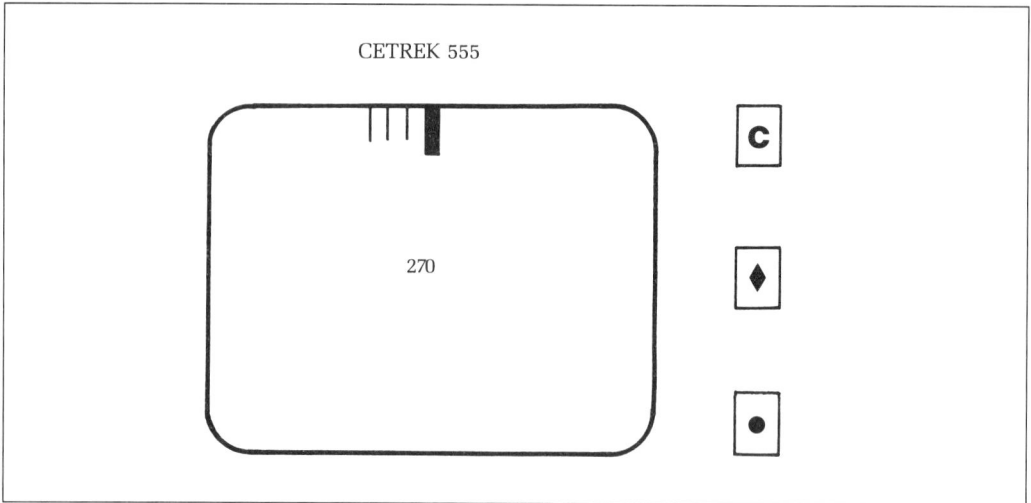

The compass in off-course mode. The boat is off course 3 degrees to port.

the trend of any change of direction and how fast it is happening.

A second part of this function fixes the digital display on the course you have selected to steer, whilst the analogue spikes show how and by how much you are deviating from it. A 1-degree error to port causes a single spike to show on the left. An error of 2 degrees means two spikes and 5 degrees lights the entire segment to 11 o'clock. An alarm can be set to sound when the boat gets further off line than the skipper is prepared to tolerate. This function is excellent on a sailing boat because it shows how much you are luffing, or are letting the bow drop off the wind. It is also a very good way of monitoring how well the autopilot is performing and helps the skipper decide if it needs a bit more anti-yaw control, or more help for the rudder.

In fast boats, I go everywhere in off-course mode. It is a remarkably efficient way of keeping the boat on course over long distances, especially in poor visi-

bility, or when you are too far offshore for any landmarks to show. As long as the screen remains clear of spikes, you know that she is on track and that you are not steering a zigzag, overcorrecting course which frequently crosses the rhumb line – as often happens to ordinary compass watchers.

**Off-course trend** is the third function which acts as a steering monitor by memorizing the helmsman's temporary deviations and computing them into a general direction line. In effect it tells him that even though he is aiming to steer 090, he is a bit right-hand down and is more often inclined to steer south of this heading than north. A skipper running a dead reckoning will consult this feature from time to time and use its information to make his chart plot.

**Head and lift** are pure sail boat functions showing the helmsman whether the wind is heading him towards and so moving him on to a poor course line, or if he is being lifted on to a better line – in other words clawing more ground towards an upwind target. The function can also be set to memorize port and starboard tacks so that the boat can very quickly be settled on to the correct heading as soon as she is brought about.

**Current trend of head** is really a big boat or ship facility. Some of these turn so slowly that the transverse movement of the bow is almost indiscernible to the eye. Captains get very anxious if they have ordered a course change but the ship's head appears not to be coming round against the wind or is coming round too fast.

In this mode, direction is again shown digitally, but a screen edge giroscope spins clockwise or anticlockwise to show turning trend. It also speeds and slows to show the rate of turn.

## Other Systems

The features described above are those on my own compasses. Certain manufacturers omit some of them and most have their own way of displaying the information. Off course, for instance, can be shown by a swinging needle, an LCD boat upright on the screen and deviating either side of head up, or it can be indicated by red and green lights illuminating on the appropriate side and in deviation ratio. Each user will vary in his preference for screens and dials and in the number of functions. There is plenty of choice.

---

**SUMMARY**

- There are several types of electronic compass.

- The flux element is a coil which responds with varying voltage to changes in the earth's lines of magnetic force relative to boat heading.

- The fluxgate compass settles in milliseconds after a rotation; this is its prime advantage.

- The sensor is installed at the centre of the boat's pitch and roll.

- The readings will be inaccurate during hard acceleration.

- Automatic deviation adjustment is very simple and very reliable.

- Automatic deviation adjustment cannot be done on land and certainly not on the boat trailer.

- A good off-course mode display is of paramount importance.

- Head and lift functions need to be learned and practised.

- Beware of tilting a hand-held electronic compass.

# 6
# THE AUTOPILOT

An autopilot is often very low down on the newcomer's list of equipment. Indeed, there are also many coastal skippers who decry automatic steering devices and who comment that they enjoy driving the boat, so to have this function usurped by a machine would take away half their pleasure.

There was a time, during my day cruiser phase when I might have had sympathy with such a parochial view, but now that I have done many long trips with Wilbur, my electronic helmsman, keeping the boat on course for hour after hour and day after day, I cannot imagine a contented cruising life without him. Despite price trends, he was not too expensive and has proved to be super value for money.

Unfortunately, some of the earlier autopilots in the leisure field were not very accurate and some were neither robust nor electrically very reliable. This syndrome tells you immediately that the efficiency of any self-steering mechanism depends on:

1.   The quality of the compass.
2.   The quality of the engineering and displays.
3.   The design of the electronics and software.

These are the things you must look for when you buy. If a good, rock-steady autopilot is a very comforting cruising aid, one which gives you doubts, or which jumps

off course, or cannot control the boat in severe gusts is a recipe for a nervous breakdown. An autopilot is only desirable if it inspires total trust. The range you will be offered is from a tiny helm pilot to a totally independent machine, which can be so integrated with other systems that it will guide the boat from place to place, over a complicated course, by using information taken from Decca or GPS, and will even use radar generated information to avoid other vessels in its path.

Once, when taking a motor yacht from the Solent to Scotland, my partner and I were imprisoned in Penzance by bad weather. Jokingly, but true, we discussed that we had enough on-board technology

A typical tiller pilot for a small yacht.

Larger boats need bigger pilots.

to send the boat around wave-battered Land's End on its own. We could have taken the train North and rejoined the unmanned ship off the Isle of Man!

Strangely in this humorous, but entirely feasible, scenario, we suspected the performance of some of the other equipment, but never for one moment doubted the ability of the automatic pilot to control that big boat, no matter what wind and waves might be thrown at it.

## So, Do You Want One?

Safety at sea should demand high quality

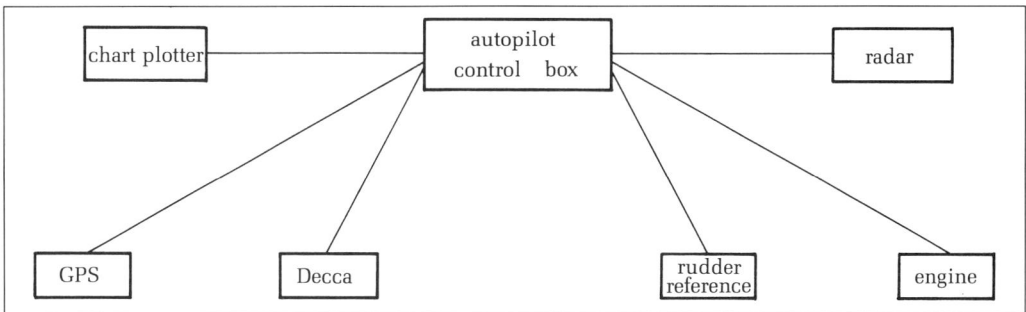

You can link the pilot to other instruments.

which creates the following very good reasons for installing a strong autopilot aboard a cruising boat:

1.  The autopilot takes all the physical fatigue out of long-distance steering, so conserving the skipper's stamina. It also means that he can keep his eye on the boat's progress from the shelter of a spray hood, or by perching himself while well braced at the head of the saloon steps.

2.  The well-adjusted autopilot will always steer the boat better than any human, especially in such adverse conditions as big waves on the quarter, or abeam.

3.  The skipper can take his eyes away from the compass more often and will be able to divide his time between chartwork and keeping a good lookout.

4.  Because of its superior course-keeping, the autopilot puts less strain on steering systems and causes less wear and tear.

5.  It can be set up as such a totally independent system that you could steer the boat on the autopilot even if the prime steering system breaks down. This last very desirable feature, however, can only be done with certain automatic pilots.

The usual range of autopilots is given below.

## The Tiller Pilot

As its name implies, this can be used on tiller-steered boats. The most common type is an oblong box containing a fluxgate compass and an electrically driven arm. The ensemble is bolted to a fixed part of the cockpit coaming at one end, whilst the other attaches to the tiller.

Most tiller activating pilots are set up by putting the boat on course, then aligning the compass dial with north, or some other indicator. When the sensor detects a deviation from north, it returns to this pole

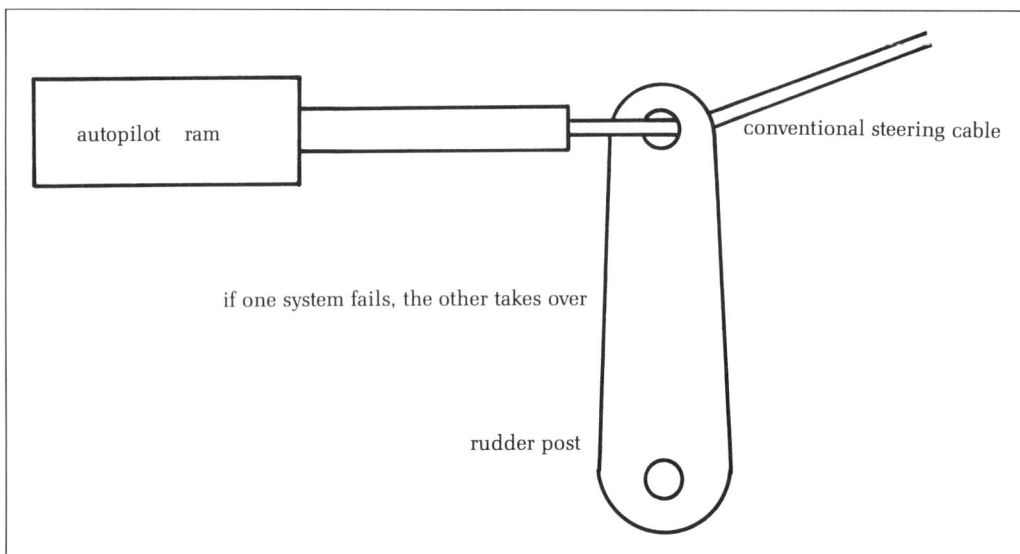

autopilot ram

conventional steering cable

if one system fails, the other takes over

rudder post

Autopilot ram straight to the tiller quadrant makes sense.

position and brings the boat's head back by the same amount.

Tiller pilots are generally very simple and very reliable in moderate weather. They have the inherent problem that the control mechanism is outside where it can be exposed to the weather. Also, if one was made powerful enough to control a large, tiller-steered yacht in any weather, it would be very heavy to lift on to its attachment, a bit bulky to stow and probably too expensive to attract small-boat owners. Happily, new materials and better electronics are making great improvements in this direction.

## The Wheel Pilot

This is also described succinctly by its type name. There are a number of mechanical configurations but they are all dependent (usually) on a fluxgate compass sending out a varying voltage which mirrors the bow's deviation from the desired course and applies enough reverse wheel to correct it again.

The main differences are in the way the driving power is connected up to the steering. A very simple, but effective method is to have a geared electric motor mounted close to the steering. This is designed to rotate clockwise or anti-clockwise according to the electrical message which it receives from the compass output. A notched drum is mounted on the back of the steering wheel and is connected to the drive unit by a toothed, rubber band (car fan belt). When the boat is set on its course, the device is tensioned by a simple lever mechanism and the autopilot takes over. Changes are made electrically, of course, either by push buttons for plus or minus a few degrees, or by rotating a dial.

In spite of their flimsy appearance, belt-driven wheel pilots are very powerful. It makes the principal manufacturer shudder, but I know several skippers of fishing trawlers weighing in excess of 10 tonnes who use such pilots which were originally designed for yachts. They even use them with the trawl down. 'Nothing to go wrong. Everything in sight. Easy to maintain', they comment.

A few such pilots use a steel cog wheel and metal motor-cycle chain to perform the same task, but they tend to be oily and noisy. They also suffer the inherent drawbacks of all belt and chain systems of having an inevitable slackness. This makes them less precise and slightly more prone to wear, because they must always take up this slack before they get into drive – but they are still very practical.

If you require a high degree of precision and immediate, fine-tuned response to electronic direction keeping signals, however, you must move up to the more sophisticated and more expensive levels of hydraulics.

## Hydraulic Autopilots

These also vary in size and form, with the most usual being some type of pump connected into the boat's normal hydraulic lines and driving, or resisting, the rudder as the compass dictates. In order to do this it needs to know exactly how the rudder is responding and at what angle it is currently set.

In order to achieve a fine degree of precision, all sophisticated automatic pilots need to refer to the rudder. Probably the most usual is some sort of potentiometer and swinging arm pivoted on the rudder stock or quadrant. As both move in

A typical wheel activating autopilot.

A heavy-duty pilot mechanism.

concert, it varies the electrical signal, which is converted into appropriate amounts of hydraulic push and pull by the electronics.

A further advantage of this system is that it can be used to display rudder angle at the helm position. This is a favourite and much used facility on my own boat, not only to show the skipper where the blade is when he is manoeuvring, but a quick look lets him set the rudder and tension the wheel, before he blasts astern and slams the rudder and cable damagingly against their stops. Once moored, this is also a quick way of ensuring that the helm is amidships before lashing the wheel.

## Variant Hydraulic Autopilots

These systems use a very heavy-duty pump, which is attached to an equally robust hydraulic ram. The outfit is bolted down where it can be made to push or pull on the steering quadrant, or against a short tiller arm, as desired. This separateness confers two further advantages:

1. It brings all the benefits of precise, drag-free automatic hydraulic steering even if the boat is hand-steered by worm drive, pulleys, cable, or chain steering.
2. If the normal steering becomes useless, the boat can be brought home and even put alongside a quay entirely on the autopilot.

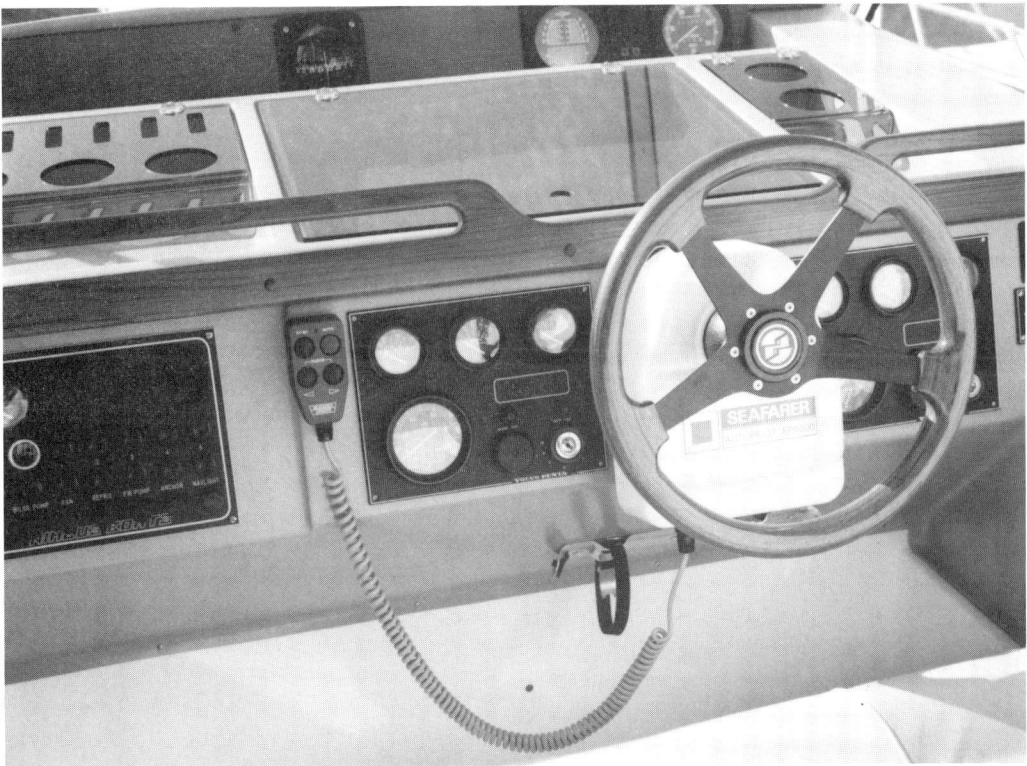

Pilots can also be steered on a remote control.

In a perfect world of unlimited finance and plenty of on-board room, we should all probably opt for the efficiency and safety of one pilot driven by the normal hydraulics, with a bypass for hand steering, with another system operating an independent ram. We should probably then also clamour for a reserve control unit.

Within these categories there are a number of autopilots designed for specific sorts of vessel. The types available cover not only differences in size, but also claim special features for planing boats, sail craft and sterndrives. Some of these claims – those for sterndrives, for example – are valid, but in the mid-price range from the established companies, you will usually find a steering device which you can move from your 3 tonne sports cruiser to your 10 tonne displacement boat without loss of efficiency. This versatility is a by-product of good electronics. Modern microchip technology has totally revolutionized the way an autopilot can subdue the two big steerage enemies and, in so doing, reduces the huge strain on itself and on the boat's other gear.

## Knowing the Enemy

### Understeering

This occurs when the boat repeatedly drifts off line to one side. This is especially the case with sailing boats and with all vessels being influenced by diagonal waves, or a current aslant the bows. It means that the control systems are bringing the boat almost back to line, but are shutting down too soon, so the head wanders again.

The remedy is to increase the amount of rudder (this forming a greater angle of deflection) applied to pull the head more vigorously back on line. It instructs the drive to work harder.

### Oversteering

This is the converse of understeering and can even by caused by applying too much correction to understeering. It means that the boat's head is passing across the line of the desired compass course, but then continues to turn a long way beyond it before correcting itself again.

In oversteer, the autopilot is behaving like a newcomer to boating who fails to 'catch' the head by taking some turn off the wheel before he reaches his line. He is not obeying the old sailing skipper's oft-heard command 'Meet her, Mister, before she goes too far.'

An oversteered boat wanders across the ocean like someone conducting a zigzag search, but the electronics can combine with the rudder reference unit and the high-speed fluxgate compass to reduce the amount of rudder applied when a slight deviation off course is sensed. It is, in effect, asking the autopilot to be a bit less brutal.

### Yaw

This is very often caused by waves on the boat's quarter. In many ways, the picture is the same as that associated with oversteer and understeer, but once the boat has been set up to control these (all boats require different settings) the settings can be left untouched. The yaw control is frequently altered to cope with changing wind, waves and weather.

The symptoms of yawing are that the boat gets a long way off course on both

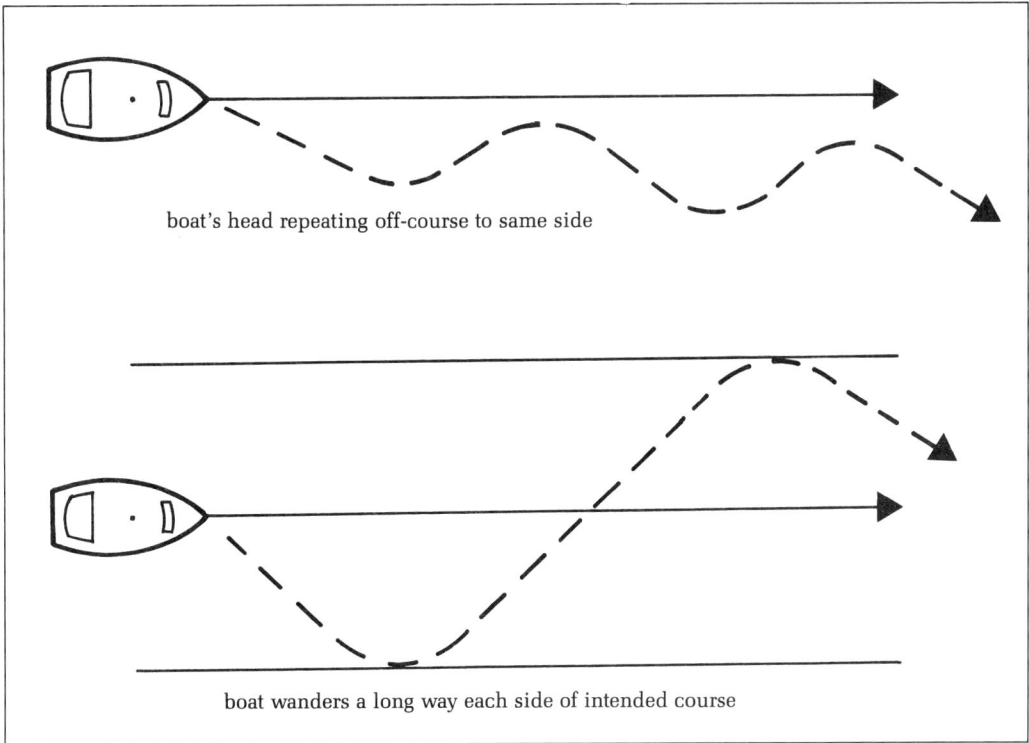

boat's head repeating off-course to same side

boat wanders a long way each side of intended course

Understeering (top) and oversteering (bottom).

sides of the rhumb line before any rudder correction is applied. In effect, the autopilot is not responding quickly enough to boat deviation. The description explains why some manufacturers actually call this the response function. It is the autopilot's mechanism to allow for weather.

In big waves, speeding up the response time will 'catch' the head more quickly and apply early counter rudder, just like a good helmsman will sense that an approaching wave on the quarter is about to accelerate the boat and to push its head up into the wind, thereby pulling the tiller towards him before it happens. We should all give much more attention to this particular function.

Because your boat does not change from day to day, the rudder setting can usually be set and then forgotten. The anti-yaw control should be verified every time you go to sea. The usual error (which we all make) is to increase this control on Monday to counteract waves, and then forget to reset it on Tuesday, when we trundle along wondering why the boat is oscillating from side to side and the wheel is spinning like a demented top.

## A Basic Set-Up

Good practice is to have a flat sea setting – in other words, a minimum – and work up

This one set of circumstances alone explains why many skippers fall out with the automatic pilot. They pay to have it installed, then expect to switch it on and have the boat under control without further ado. Life is not that simple. The workings of an autopilot have to be learned in conjunction with the boat it drives, just like with any other piece of equipment. In this there is no substitute for experience. I have been running the same autopilot for three years and am still acquiring more small tricks for better use. Most of these will be employed on any trip of more than 10 miles.

The useful rudder reference display.

## Finisterre to Camarinas

To illustrate some of the tricks I have learnt through using my autopilot, I shall describe a journey I made with my partner from Finisterre to the Ria de Camarinas in north-western Spain.

We decided to leave at first light to avoid the gusting wind which occurs during

from there. If in doubt, reset to the low point and tweak the setting upwards gradually until the boat holds her true line.

Like most boats we have dozens of switches.

most afternoons in this area. We did the nav planning the night before, put all the waypoints into the two electronic navigators and marked the course on the chart. Both black boxes and the manual calculation showed an expected trip of about 20 miles, around seven waypoints, plus a run down the leading marks into the deep and tricky Ria de Camarinas.

At 06.30 we flashed up the diesel on a flat calm morning, then went through the ritual of closing the dozen switches to activate all the instruments, two of them belonging to Wilbur, our automatic pilot. One is a general light-duty supply, pulling a couple of amps at most, the other is the current to the hydraulic ram which can draw up to 25amp for a millisecond when it is really working for its living. Normally the ammeter needle deflects to about 10amp for a second, just to remind us that the autopilot is doing its job.

All the LCDs came on as Wilbur went through his self-test routine, finalizing with the message that the boat's head was on 225 mag, pointing into the corner of Finisterre's tiny crowded harbour. The fluxgate compass repeater showed the same and everything else looked fine, so we pulled up the anchor. We have a code of hand signals from hauler to helm to make this easier. Before driving the boat yet further up to the big CQR, whoever is on the wheel glances down at the pilot's rudder deflection indicator and sets the wheel accordingly.

Up came the anchor – full of mud and weed – which we stowed temporarily, and then we joined the dozens of flared bow Spanish fishing boats whose crews gave us a cheery wave and called 'Hola' as they overtook us on the way out to their day's work. At this point we were hand steering to get around the harbour wall on the first leg of a complicated route which would take us south down the east side of the notorious Cape Finisterre, around its rock-strewn foot and up north again towards Camarinas.

As though to prove the folly of total reliance on electronics, one of our two navigator boxes would not lock on to the first waypoint just 3 miles ahead. It took us a while to realize that waypoint four was actually only 2 miles due west as the crow flies across the Finisterre Peninsula. The computer worked out that because WP4 was closer than all other marks, we must have passed them already. We put Wilbur on to our manually calculated course, set low revs, checked that he had the boat under good control, then both went forward to sort out the anchor chain, lash down the bikes and take off the sail covers. By now we had earned our second cup of tea, and although Wilbur cannot actually brew it, he frees one of us to do it.

The GPS/Decca alarm 'bleeped' to say that we were 0.1 nautical miles from waypoint one and had been pushed 300m (500yd) east of our track by the current. We decided to come round early and to do so without disengaging the pilot. Because autopilots do not like violent changes of course, we made the 25-degree shift to starboard 5 degrees at a time over a couple of minutes. Such a change is very simple. We merely needed to rotate the control button whilst watching the autopilot's digital read-out of heading. It is also comforting to watch the green LCD light up to show that the pilot is responding correctly to a change of course to starboard and to see the steering wheel confirm this with a clockwise twitch.

Along this next leg, my partner spent a lot of time glancing between Wilbur, the

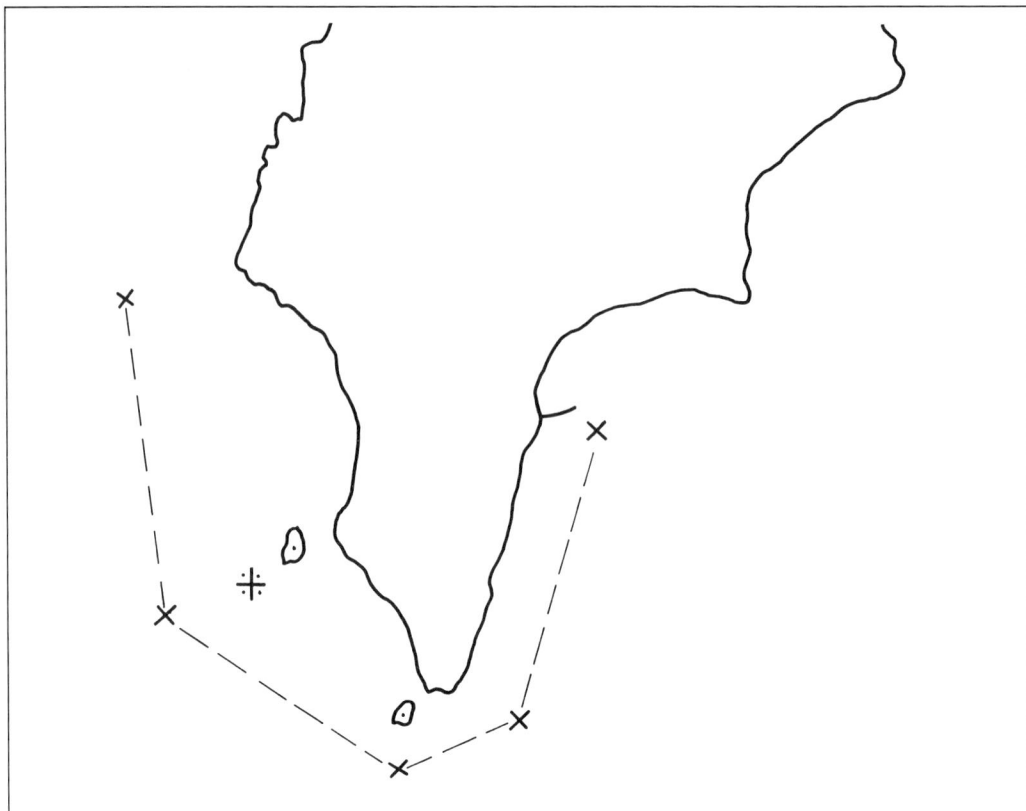

The route around Cape Finisterre.

bow and the main steering compass. 'He's not happy' she decided. 'He is hanging left too much.' We discussed the big, but gentle Atlantic swell and the ebb coming around the corner and decided to leave well alone for the present 2 mile leg and look at it again when we would be settled on the 8 mile stretch which was to come next.

'He's not happy on this leg either. The head is getting a long way out to starboard before coming back, but does not do the same to port. We had best have a look at the settings.'

Sure enough, as a compulsive button pusher, I had been playing again and the rudder and response were set too high. As soon as we settled them, we also settled the boat and plodded 5 knots north with a catspaw of wind developing. Wilbur kept the boat whilst we both went up on deck to get on the sails – with one eye open for lobster pots and trawlers. We were going motor sailing.

As soon as we had all sails on, the big genoa put 1.5 knots on the speed and the strain came off the engine, so the revs increased a couple of hundred of their own volition. We have a left-hand prop, which tends to 'walk' the stern out to port. The pilot was now struggling a bit to counteract this, so we tweaked up the response to cope, gave it five minutes to assure ourselves that the head is settled, then decided on coffee.

The next leg involved quite a sharp turn, so we used 'method two' to get on to it. With the boat being steered by hand, we set Wilbur's control panel to the next course to be steered. Then, when we were within 10 degrees of it, we pushed down the button marked 'pilot', got on to electronic steering and had the satisfaction of seeing the boat's head nod very slightly beyond the setting, then immediately lock on again and hold the straight line.

This bit of coast is absolutely strewn with lobster pot markers lurking to get caught up in an unwatchful boat's propeller. For safety's sake we unrolled the remote control lead and took the control panel across to that wheel-house window which gives the best forward view. Sure enough, twice in the next hour we almost ran down a marker buoy, but were able to put in a 'dodge'. As long as the operator holds down the appropriately coloured button, the boat will continue to turn to that side. As soon as it is released, she comes back on to her original course again. In Lobster Pot Alley it is a very useful device, saving a lot of wheel time and the fiddly business of putting the boat manually back on an approximate course then refining it with the fluxgate and autopilot.

## CLOSE CONFINED

The Ria de Camarinas is not my favourite. It has too many off-lying growler rocks and shoal patches. Avoiding them means steering some very precise lines. With a south-westerly gale forecast for Finisterre, we did not want to be playing around in wind and waves, whilst also wondering if we were in the right place.

In this sort of situation, modern equipment is superb. By using it in combination, we knew that we should be able to hit a mark at the head of the Ria in an uncertainty circle of no more than 100m (100 yd), then be able to see the leading marks on the shore and follow their indicated line into the calm water.

Over the final mile, we watched the instruments very carefully. The magnetic

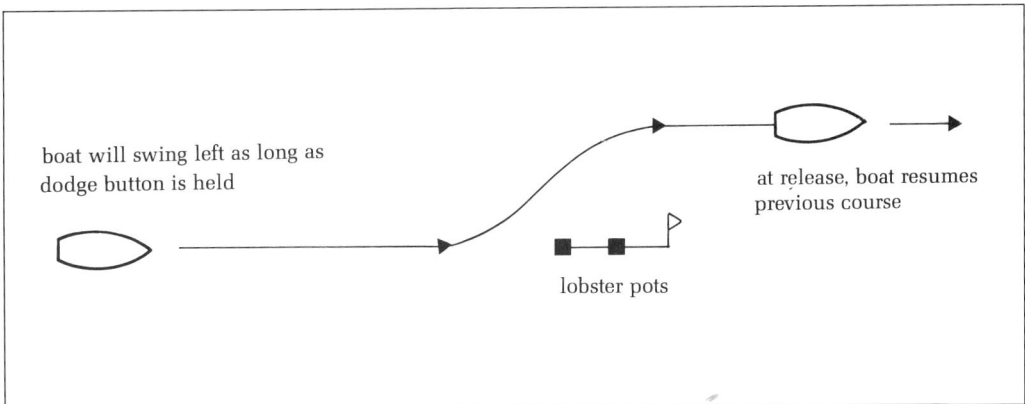

boat will swing left as long as dodge button is held

at release, boat resumes previous course

lobster pots

How a dodge works.

compass heading to our mark was 040. Wilbur agreed that he was holding the boat's nose exactly where we wanted it. The GPS navigator and the Decca both confirmed that a current was pushing us 4 degrees left and that our track over the ground, or CMG, was actually 045 degrees. Normally we do not fight the tide too much, but right then we needed real precision and knew that electronic navigation would give it to us. We rotated

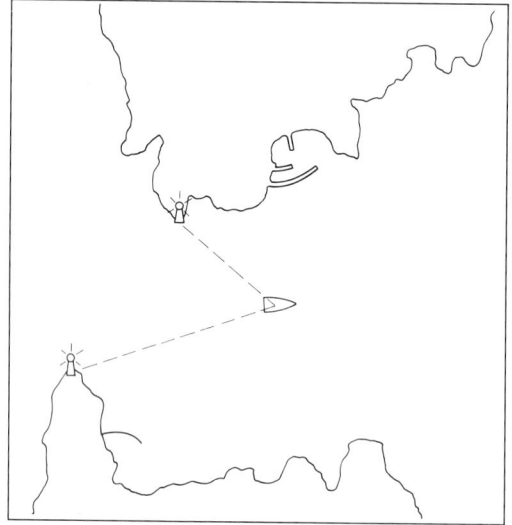

Cross-bearing on the two lighthouses to confirm the position.

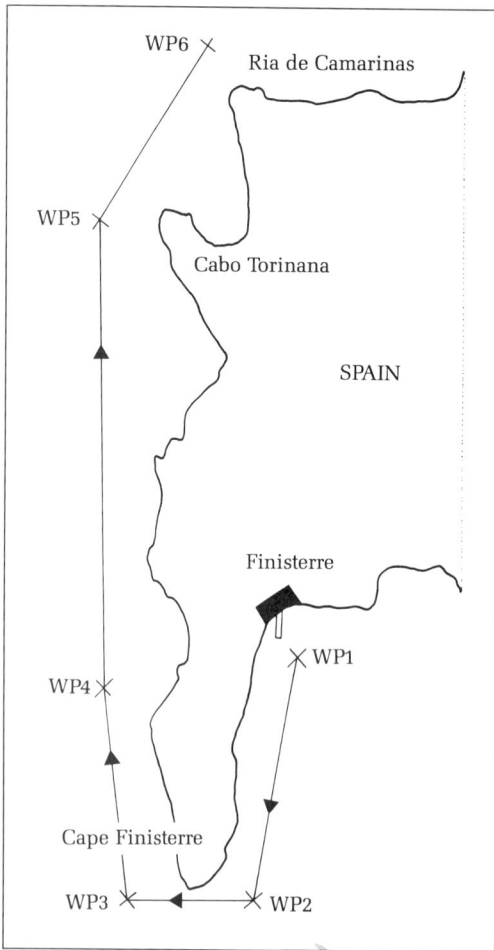

Finisterre to Ria de Camarinas on autopilot.

Wilbur's control wheel so that the readout showed 053 degrees and within thirty seconds the navigator screen confirmed that we were now tracking 049 and holding it. Along here, I had a lighthouse tower on Punta de la Barca and the huge lighthouse on Cabo Villano plainly in view, so I could cross-check the position with hand bearings and use all this information to verify that we were not drifting off our line.

## WE MADE IT

We were actually very proud of that leg. When the GPS bleeped to announce that we had reached the waypoint, I picked up the binoculars and immediately spotted the two white towers of the leading marks on Punta de Largo in line on 108T and giving a safe course into the Ria. We turned on to the line and let the pilot take us in to lunch. In the calmer water between the headlands, he held the boat steady whilst we took off the sails and showed us where the rudder was as we manoeuvred alongside a massively high wall in a wind which was becoming increasingly unfriendly.

That one inter-waypoint passage possibly explains why I remain a fan of electronic navigation and big, electronically controlled autopilots. I could not have matched that precision with less automated methods.

## SUMMARY

- Buy the most powerful autopilot you can afford.

- The autopilot's performance is governed by the quality of the compass.

- If possible opt for independence from the boat's main steering system.

- Watch out for understeer and oversteer.

- Activate all the secondary functions and learn them in calm conditions.

- Keep a constant check on the behaviour of the boat's head even when the autopilot is engaged.

- Note the pilot sensor readings and the actual heading (as shown by a good compass) and then write them in the log.

- Constantly check the autopilot heading against the boat's best compass.

- Install the autopilot controls where the off button can be easily reached – even in the dark.

- The danger of an autopilot is that the crew do not look up and ahead often enough.

# 7
# ECHO-SOUNDER
# NAVIGATION

If there are sensations worse than not knowing where you are and exactly what and how much you have under the keel, I do not want to know about them. These two navigators' nightmares really go together and explain why a good echo-sounder is as much a navigation and position-fixing device as a pure indicator of water depth. If you can see what is on the sea-bed and refer it to the chart, you

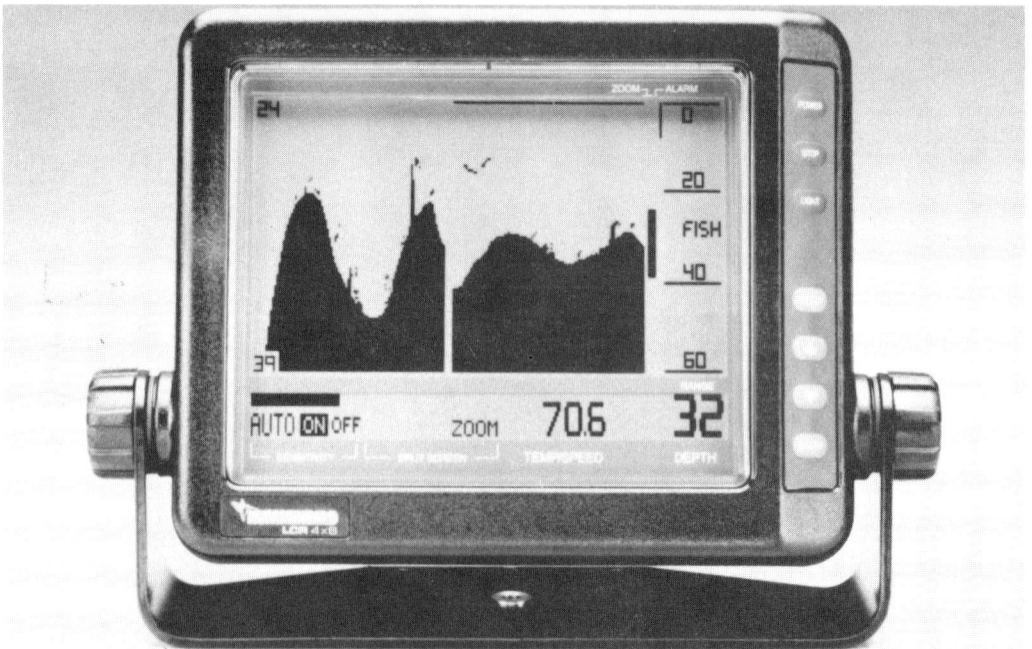

The best sounders give a view like an underwater camera.

pulses fan out
into a cone pattern

100ft

field of view is
approx. 33ft across

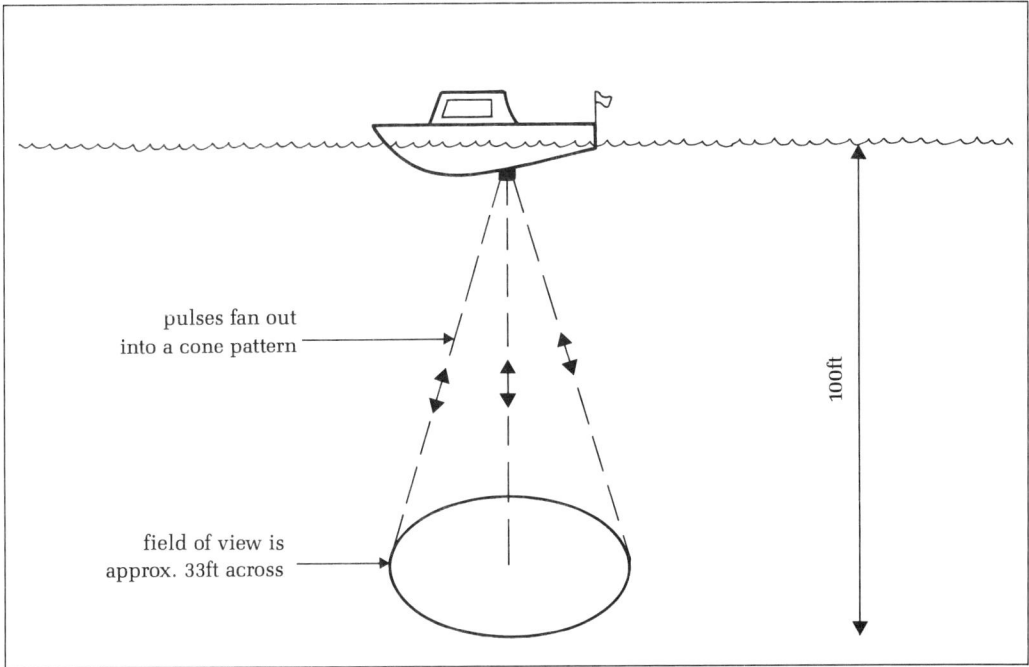

The pulses go to the bottom and are reflected back.

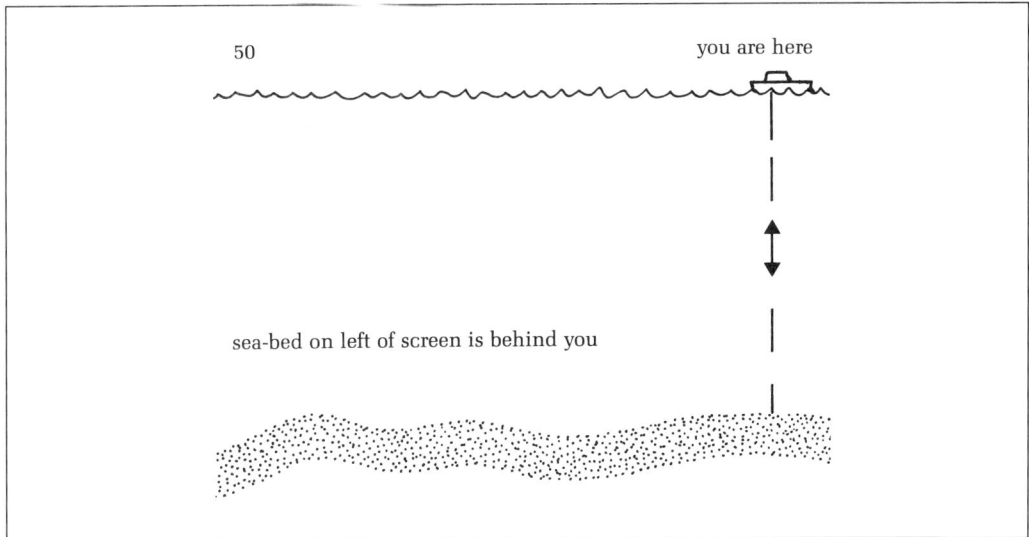

50

you are here

sea-bed on left of screen is behind you

To avoid confusion, remember that your boat is in the top-right corner of the screen.

often get a second opinion on where you are. This especially applies to those sounders which have kept pace with electronic progress. Using the best of them is almost as good as having an underwater camera.

Before I go on to discuss how they work, it will be just as well to sort out the confusion of what the various types of machine are called and to translate the engineers' shorthand into plain English.

## Basic Sounder Principles

The actual principles used to measure depth electronically have not greatly changed in the past twenty years. They are well documented and generally well enough understood not to need detail here. Suffice it to say that they are common to all machines from those which are reasonably priced to the all-singing, all-dancing breed which are very expensive. They all work on sonar phenomena, whereby an electronic signal is converted into an ultrasonic signal by the transducer, which also shoots it at the seabed. The transducer is a bit like a combined transmitter and receiver. Having sent the signal out, it receives the returned echo and computes the time taken in transit, then electronically converts this into digits, or into pictures.

In loose parlance, we tend to talk of echo sounders, depth gauges, fishfinders and sonar. They are all really the same, but the range of terminology shows that there are different machines for different purposes, but that none of them can escape the basics above. The type of machine mostly seen on leisure boats and small inshore fishing boats will be one of those below.

## *Flashing LEDs*

The Flashing LED (light emitting diode) has been around for such a long time that it is surprising to find it still in production. Its operating system is entirely normal, but its display comprises a strip of light, or series of lights, which increase in brightness and number as the depth increases. The weakness of the breed is that it is not very precise and usually needs scales to be interpreted for each of its separate depth ranges. This is not always easy to do when the user is tired, or beginning to panic.

## *Digital Sounders*

These are the next up the simplicity scale. It takes the raw information and electronically converts it to numbers displayed on an LCD (liquid crystal display) screen. The numbers are easy to read and most digitals offer the read-out in feet, metres, or fathoms as the user chooses.

Digital sounders are obviously ideal for incorporation into combined speed and depth duos for those who only need information on depth, without any

Digital sounders are now more common.

interpretation of the nature of the sea-bed beneath.

## Paper Recorders

Paper recorders are the most mature of those echo-sounders which give an idea of the sea-bed's shape and contours. The new breed does this with a surprising amount of detail and fine line-drawings. The weakness on some is again the need to interpret scales to calculate the actual depth and they all suffer the drawbacks of high running costs to replace the rolls of special paper.

Most skippers using paper recorder sounders do not leave them running all the time. There are obvious dangers and drawbacks to this.

## Video LCD Depth Sounders

These are the simplest of those which actually draw a picture of the ocean floor's

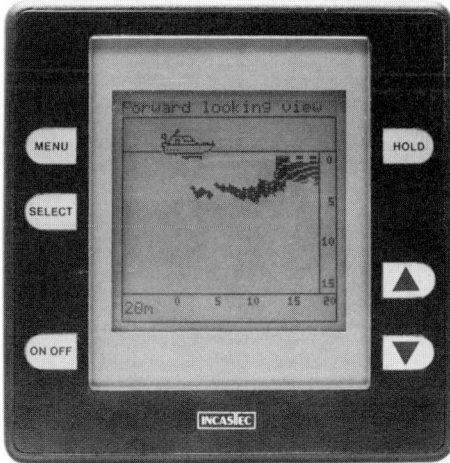

Monochrome LCD video sounder.

vertical and horizontal contours and displays them on a screen. In addition to showing digital depth, mirrored by a bar chart at the edge, the display produces a two-dimensional image of the shape of the sea-bed over the transducer's area of coverage.

The advantages of this sort of information are numerous. I use mine especially when coming up to a strange anchorage in order to determine if I have flat mud, or rocks and weed where I plan to drop the hook. Divers and anglers looking for rocks and wrecks are well served by the picture, which not only shows the wreck and gives its height from the sea-bed, but also shows whether the boat is travelling across its beam or down its length.

The LCD video's minus qualities are the coarseness of the image, which is made up of numerous square pixels and the limits which this imposes on the size of the screen. Just like digital watches, the black crystal display is not easy to read in certain lights and from particular angles. With a little practice, however, it is possible to interpret whether the bottom is rock or mud, but this is considerably easier with a colour picture.

## CRT Fishfinder

The CRT (cathode ray tube) fishfinder uses television technology to produce a very sharp image in colours that are decided by the engineer. Different substances create different on-screen hues and shades, which are very easy to interpret and even the type of fish can be guessed at by an experienced operator. Most of the latter are dedicated sea anglers and professional fishermen.

The main minuses of CRT sounders are

Super screen, but what is the amperage needed?

that even though they can offer a large screen, they are very bulky top to bottom and even worse front to back. Only rarely can you flush mount a colour sounder and even the smallest of them takes up a considerable amount of wheel-house space. Bright pictures connote big power consumption. CRT sounders are very voltage hungry, which is fine if your engine is always running, but they are hardly suitable for a sailing yacht.

## Sounder Power

Power is one of the contentious points of echo-sounder marketing. The power itself can be discussed in two forms.

## The Current Drain

The current drain of any model can be matched to the size of your boat's battery bank. You choose to suit, with a rule of thumb which dictates that, generally speaking, the black and white LCD type will consume very little (probably a few milliamps), the cathode ray tube the most and with the others falling somewhere in between.

## Output Power

The bone of contention concerns the power radiated by the transducer – that combined transmitter and receiver which sends out the signals and collects the echoes. The greater the output power, the greater the depth which can be monitored and the brighter, sharper and clearer the on-screen display can be made. Unfortunately, however, there is no common trade language for quoting output power in advertisements. Some manufacturers list the total power produced by the transmitter, whilst others give it as a root mean square (rms) figure. This is probably a bit fairer.

Raw power is a bit like trying to use a stereo radio speaker on full blast. It causes reverberations and extra noises which means that the human ear is unable to make use of the output. Echo-sounders are the same. The power needs to be controlled if it is to be of any use. If you are offered a sounder which has a very flattering quoted output at a seemingly low price, you will need to check the units – watts output or watts rms. To compare like with like, divide the raw power by 8 to get the usable rms figure.

## Echo-Sounder Language

With very few exceptions, echo-sounder manufacturers use the same language and cryptic terminology for all the things which sounders can be made to do, whether they are automatic, manual or accessed from computer-style menu screens.

### *Depth Range*

This is one function which can take plenty of operator time, or none at all. Many users are content to leave the control set to automatic, which causes the sounder to make its own range changes as the water depth becomes deeper or shallows off. In auto mode, when the electronics get a 'no bottom' reading, they will move the software to a deeper setting to show the bottom where it should be on the screen. This will not be apparent with a purely digital machine, but the video versions

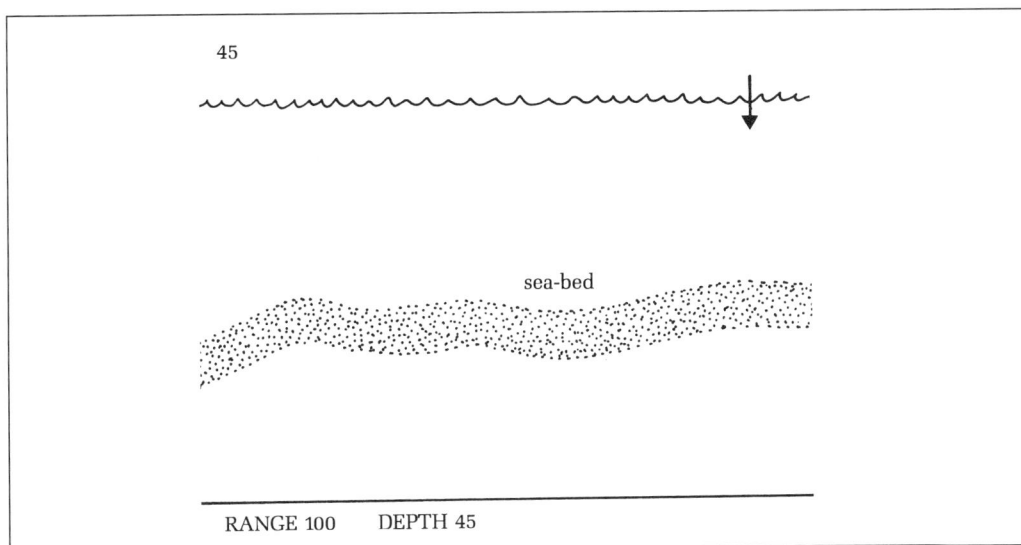

Here the range is set too deep. The sea-bed is not at the bottom of the screen.

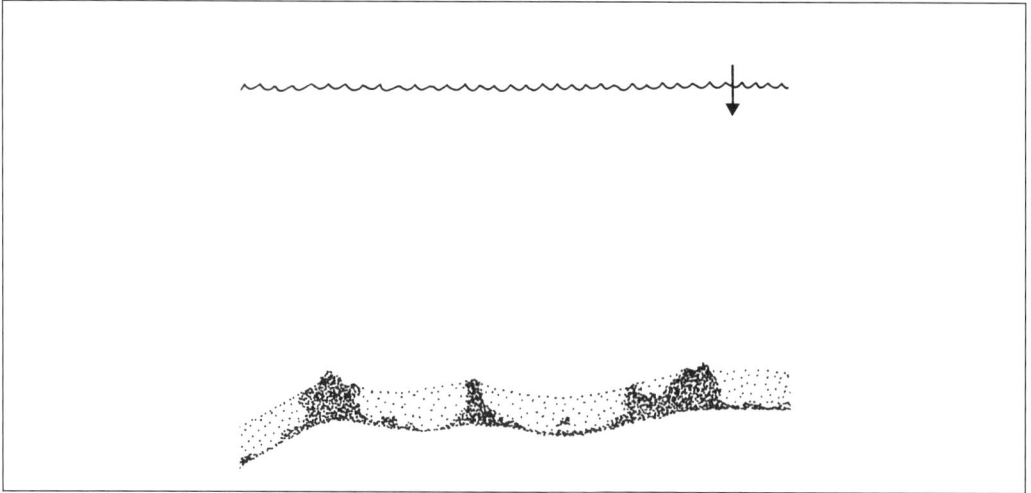

Here the bottom is a mixture of rock (dark) and sand (white).

can produce some interesting display characteristics.

With a sounder set to its 50m (165ft) range, the sea-bed should show itself at the very bottom of the screen when the water equals this depth. At 25m (82ft), it should show half-way up the screen. Very sophisticated echo-sounders will also automatically change the frequency of the sonar emissions, using a slower, heavier pulse in deep water and vice versa.

Divers and anglers often use the video sounder in manual mode. In this, the bottom is not totally coloured in as a solid mass, but appears as a double line effect which can be used to interpret the make-up of the sea-bed. The width of the second line is the indicator. If it is narrow, you can generally expect sand or mud. When a dark, wide second echo line shows beneath the top one, you are probably on rock or shingle. Sand and rocks mixed give a very informative mottled effect.

Some echo-sounders permit the setting of a 'window', in other words, you can ask them to display what is between, say 0–20m (0–65ft) or 0–40m (0–130ft). This function is a favourite amongst fishermen hunting particular species known to live at certain depths.

## The Fish Alarm

This is essential for fishing. Ultrasonic waves are greatly affected by passing through air. They will, therefore, react to that tiny amount of air in the fish's swim bladder and their change in speed can be set to trigger a buzzer in the sounder and even to display the picture of a fish at the appropriate depth point on the screen.

## Other Alarms

Other alarms are normally incorporated as 'Deep' and 'Shallow'. The operator can set the alarm to sound if the depth drops below, say 3m (10ft), which obviously has attractions for creek crawlers and other navigators. It can also be set to let the skipper know that the depth has increased beyond the limit set on the keyboard. If the route takes you over a trench or gully, the sounder will let you know that you have arrived. These two alarms are sometimes called anchor watch and are set to tell the crew that they are about to feel the keel hit the sand, or that the boat has dragged its anchor into deeper water.

Gain and sensitivity are probably best left on their automatic settings unless you are very experienced. They can be likened to the volume and tune controls of a radio. The one increases the raw power employed by the sounder, whilst the other fine tunes the signal into perfect clarity. This is occasionally useful if it is essential that you have a video picture precise enough for best interpretation of the bottom, or if you are hunting very small fish.

Display speed will probably be changed by us all at some time. Much will depend on the speed of the boat and how you wish the screen to react. Explained simplistically, a fast display speed will give a faster update on the display, but a slower speed will give much more detail.

Bottom lock is a de luxe feature found only on the best equipment. Its name describes what it does in ironing out the on-screen bumps which occur as the boat pitches and rolls. They show up a bit like small rocks and in addition to upsetting the accuracy of the recorded depth, they make interpretation virtually impossible. A bottom lock function removes wave effects so well that an angler will on occasions even be able to see a flat fish gliding along the sand.

Pixels are the separate dots making up an LCD screen, whose size is generally quoted by the number of pixels in each line and column. At other times the figures are multiplied together – generally to hide the fact that the screen is small, or a peculiar shape. The important thing that you need to know is the number of pixels per square centimetre. The greater their density, the finer will be the screen detail. The degree of contrast is also important. This can be changed electronically and by back lighting to make the screen clearer in changing light and from different angles.

A simulator is built into some echo-sounder software to enable the user to test run the gear when the boat is on shore, or to learn its functions at home.

Memory pages are also good. They store a number of wrecks, or scenes, which can later be copied to tracing paper.

# Transducers

When we think of echo-sounders, we all concentrate on the screen, whereas the efficiency of the whole unit is really decided by the quality and installation of the component which sends out and receives the signal: the transducer. These pulses radiate as a cone – a bit like a torch beam – covering a greater area as they get further from source. Some companies offer a choice of angles to suit your normal use, but only the very best can switch width of coverage by button pushing.

Transducer installation will also decide what type of transceiver device you will buy and where it can be put in the boat.

## Through-Hull Transducers

These are the most usual and the most effective. The radiation surface must be kept clear of those few millimetres of disturbed water which flow along the boat's skin (generally by a fairing block)

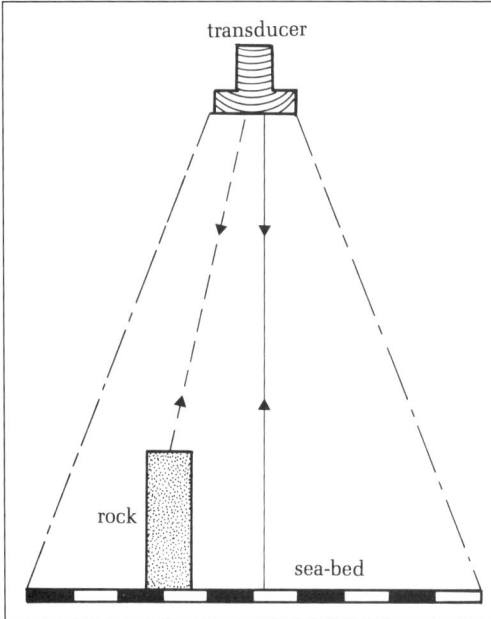

Pulse takes less time down and back to rock than the deeper sea-bed

and into clear water. Such an installation obviously involves the painful task of cutting a hole in the hull, but having the radiator in clear water makes best sense. It also means that you can add a sea temperature feature.

A transducer will grow weed and barnacles just like any other surface. It can be antifouled, but better still it should be withdrawn and cleaned. Pulling out the transducer with the boat afloat, lets in a surprisingly small amount of water at very low gusher pressure. The operation can be done without apprehension by inserting the core plug which is generally provided with the model.

## Transom-Mount Transducers

These bring many of the same benefits as through-hull transducers, plus the bonus of being simpler to install on many boats.

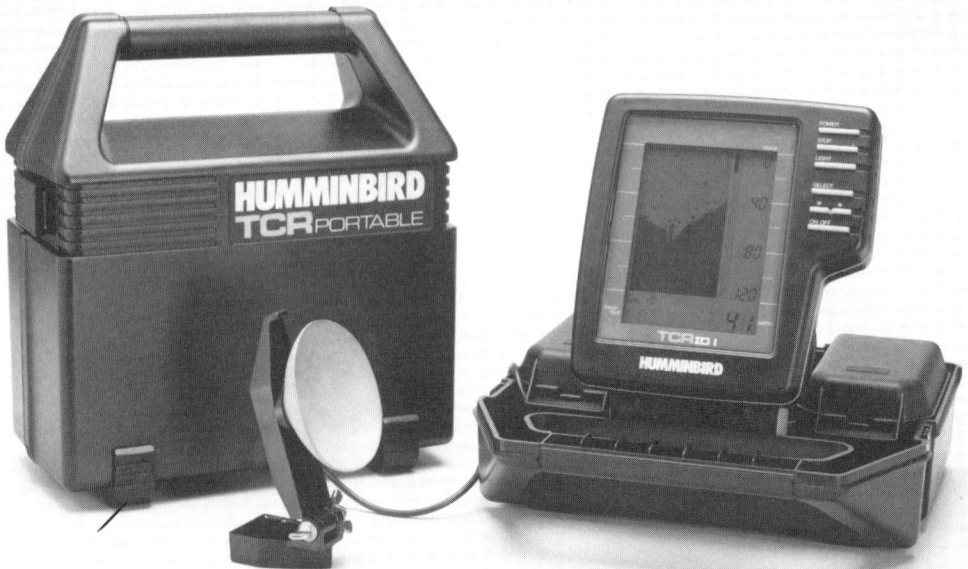

The transom transducer is a different shape.

The offset is that sensitivity might be reduced by the disturbed water which is always at the transom. Many transducers mounted clear of the boat's after line go haywire at speed and completely daft when you go astern. At slow speed they are fine.

Another consideration must be vulnerability – especially if the boat is beached, or put on a trailer. A swing-up, or spring loaded kick-up model is a possible answer.

## In-Hull Transducers

These actually have less minus qualities than first impressions imply. There must obviously be some loss of sensitivity and the actual location needs to be chosen so that the unit really is viewing vertically and not through masses of hull fittings or ballast. The great transducer enemy is air, so it must be kept away from bubbles and spaces.

The usual installation method is to make some sort of bath, or small reservoir and fill it with oil. This is viscous enough to be bubble-free and to resist too much movement. Purists insist on castor oil, but in practice, ordinary light-duty engine oil does just as well. In fact, I currently have a video echo-sounder working perfectly happily, with what is really a transom transducer, sitting in a bilge space into which I deliberately put an inch or so of water. I also once had one which was bonded carefully to the hull with a thick layer of glass resin.

## The Essential Facts

All this information shows that an echo-sounder is a paradox – a navigation instrument which is both precise and imprecise, and wherein most of the imprecision is introduced by the installation and by the operator.

The point from where the transducer actually shoots the impulse will rarely actually be at the lowest point of the keel. My own boat actually draws 1.4m (4.6ft), but she goes aground when the external transducer reads 0.9m (3ft) and the in-hull unit reads 1.2m (4ft). You can, of course, calculate these facts, but very few sounders can be calibrated to the last inch or so in 1.8m (6ft) of water, so knowing your boat and its equipment well, from observation and experience, is more use

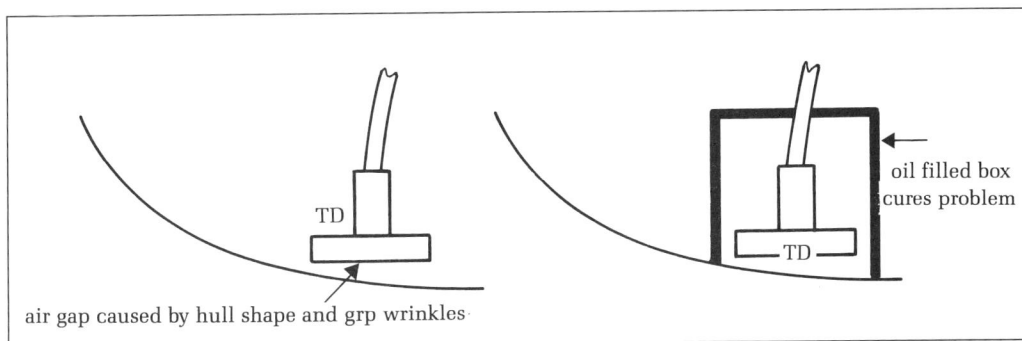

air gap caused by hull shape and grp wrinkles

TD

oil filled box cures problem

TD

In-hull transducer needs protection from air.

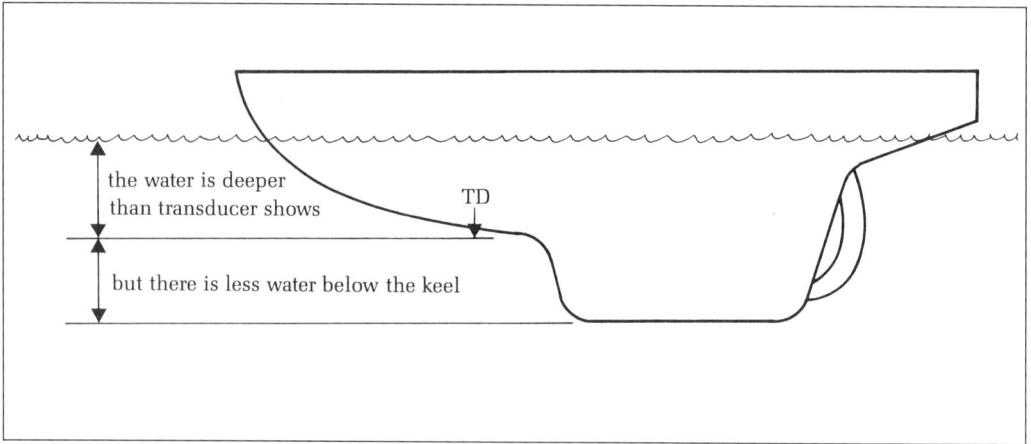

the water is deeper
than transducer shows

TD

but there is less water below the keel

The readings must be adjusted for water-line.

than pure theory. If you dry out, note the depth as you come afloat. If not, put a plumb bob over the side in flat, shallow water and compare its length to the water-line with your sounder reading.

Knowing your precise grounding depth is undoubtedly the most important navigational and safety fact of them all. It is not the sea which the sailor fears, but hitting the land. Echo-sounders are most useful when you are in shallow water. Creeping up to our own moorings in a tidal estuary is a prime example. The channel constantly changes, so shore marks are useless. I always approach on echo-sounder trend, and when the readings come down, I steer the other way to get back in mid channel again.

There are some clever writer-navigators who claim that they can use an echo-sounder as a prime passage-making tool. I admit that I am not that good. To me, the sounder is sometimes essential very close to land and for anchoring. At sea, it is a very comforting backup, most useful not so much for telling me what is under the keel but for reassuring me by confirming what is not there.

In theory you should be able to navigate along slowly in fog at, say 5m (16.5ft) depth contour as shown on the chart. In practice, this does not really work very well (particularly in waves) because only rarely will you have sufficiently precise information about the exact tidal height for that particular day, range, place, variation and barometric pressure. Practical echo-sounder pilotage is at a much lower level than its theory and has been pushed even further down the importance league by other electronics. Having said that, I often ask the other watchkeeper to note if I pass inside, say, the 20m (66ft) line and constantly monitor chart and sounder for shallow spits, rocks and other bottom features which plot the position on the chart with visual ease and certainty. When crossing the Channel, it is always interesting to see when the enormously deep trench of the Hurd Deep shows up on the screen.

That neatly sums it up. An echo-sounder

would always be one of my first electronics purchases for safety reasons. Both digital and video are useful. Even though I do not use it all the time, I am constantly glancing at it when at sea and permanently watching it as I make landfalls or go into harbours and anchorages. A good sounder is one tool I do not like to be at sea without.

### SUMMARY

- Various sounders have different applications, make sure that yours will do the job you want.

- The greater the output power, the greater the operational depth and the more accurate the readings, but the bigger the purchase cheque.

- Verify output power terminology – RMS versus total output.

- Even an expensive sounder will need calibration against an exact known depth – for example, a string and weight in the marina.

- Dual frequency sounders cover a greater depth range.

- Sounder controls and language will need to be learnt just like those of any other marine tool.

- A fast display speed is not related to boat speed but to the amount of display detail required.

- Think about transducers even before you think about displays.

- Learn the relationship between your own harbour approaches and the *en route* sounder readings to ensure that you get home safe.

# 8
# ALL THE EXTRAS

Were I to be pushed into the folly of describing the 'compleat' electronically aided human navigator, the definition would include two prime considerations:

1. He must have a boat which does not rely on just one system (say, Decca) but should equip himself with plenty of gear to do all the things necessary for getting around the world.
2. He must be able to interpret the data so that he knows precisely where the ship is at that moment, and where it can safely and speedily go next to reach its destination.

Navigation is information. You cannot have too much of it and must use it absolutely all the time. Then it not only makes your passage safe, because you are 'on top of the job', but it is also very satisfactory and lots of fun. The 'compleat' navigator draws together all the information strings from his various sources and also knows that he has others as back-ups, when he needs them.

## VHF Radio

The VHF Radio is probably the most used and least understood of all on-board systems. Newcomers to boating are confused by two facts: licensing and usage; and range. (It also has to be admitted that many very experienced seagoing people who use VHF radio very frequently also have little idea about these same things.) Let us take them in turn.

### Licensing and Usage

A VHF licence is required by anyone who wishes to talk on the system. I am not discussing ham radios, or citizens bands here, but am speaking about the normal very high frequency, shortish range transmitter found on most inshore and cruising boats. The VHF radio is sometimes misnamed as the radio-telephone, or ship-to-shore; both of these are wrong!

The exception is that any person can use a VHF transmitter, so long as they are under the supervision of a licence holder. This lifelong VHF certificate of competence is separate from the ship's annual licence and different from yet another piece of paper authorizing the use of a portable set. In practice, getting the licence is quite simple. A one-day course culminating in a not-too-painful written and practical test suffices for most people.

Out at sea, VHF comes into its own in a number of ancillary ways. Once the novelty has worn off, it has much more use as a purely receiving device rather than as a transmitter. Only on very long journeys away from the coast is it necessary to bother the rescue services with news of your departure, and I only do a radio check

ABEMA'S INFO PANELS

| VHF radio | Log | Wind | GPS |
| --- | --- | --- | --- |

| Video Depth | GPS | MF Radio |
| --- | --- | --- |

Digital Compass

Radar

Compass

| Clock | Decca | Navtex | Circuit Monitor |
| --- | --- | --- | --- |

| Eng Hrs | RPM | Temp FWD | Temp AFT | AMPS | Rudder Angle | Barograph |
| --- | --- | --- | --- | --- | --- | --- |

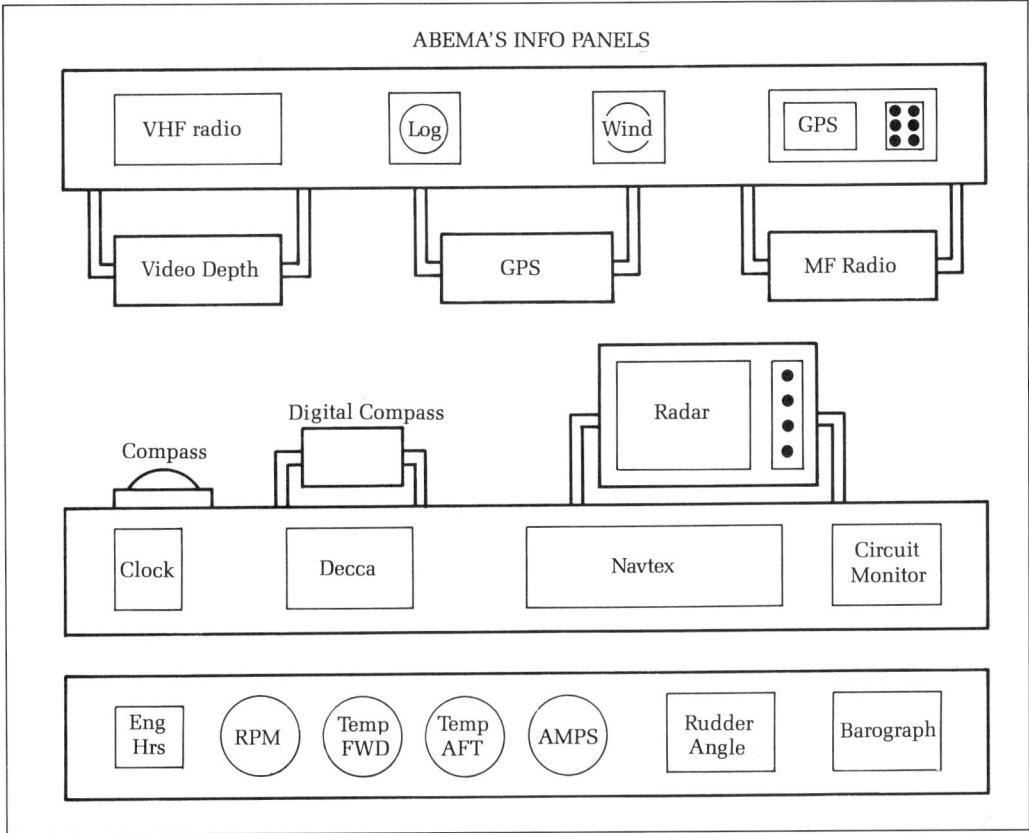

Schema of all the sources of information.

with a club-mate once at the beginning of the season.

The telephone call linked into the land system via a coast radio station is useful, provided that you have set up the arrangements about paying the bill. Marinas like you to call in to book a berth. These days, only rarely do I talk to other boats, but I listen all the time. The radio directional lighthouses, for instance, are an interesting novelty. On the other hand, monitoring the Channel 16 international calling and distress frequency is obligatory and wise. I also get plenty of information on ship movements, on what I am likely to meet in narrow channels, where yacht and power-boat races are happening, which berths are getting crowded and even where fish are being caught, just by listening in the right places. Above all, I get good weather information both from HM Coastguard and from the various other coastal authorities who broadcast regular present-state and weather to come transmissions. All these factors affect where you will navigate your boat and under what conditions. Weather is the most frequent topic of conversation amongst

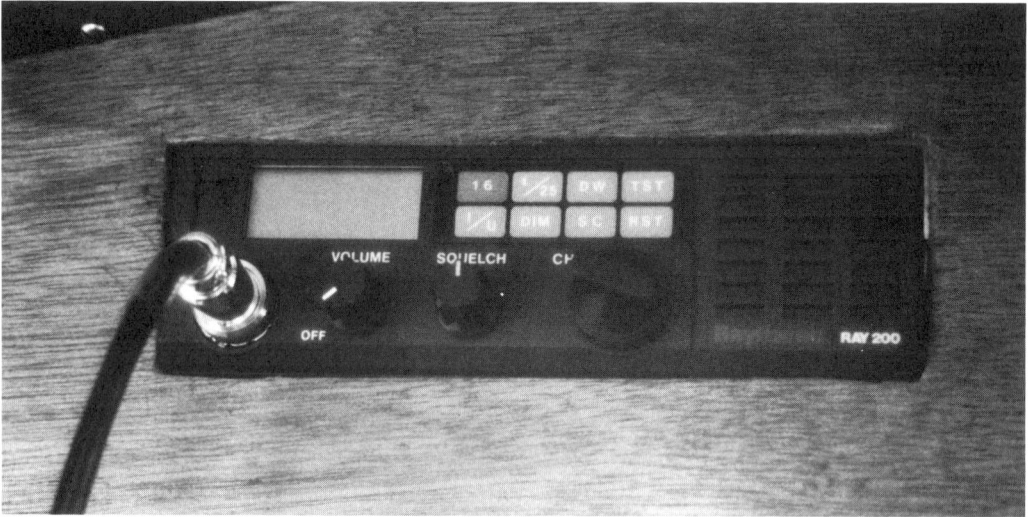

A fixed rig gives 25 watts output.

people who go to sea, but it is not generally realized that it is also a very important ingredient in how far a radio signal will travel.

Super hand-held VHF, but still only 5 watts output.

## Range

Radio range preoccupies many new boaters. It is decided by four principal factors: output power; antenna efficiency; antenna height (at both ends); and propagation conditions

### OUTPUT POWER

On the marine bands output power is limited to 25 watts. Most sets will switch between this maximum and either a 5 watt or 1 watt low-power setting which should be used for short-distance communication. Hand-held radios are mostly 5 watts and 1 watt.

### ANTENNA EFFICIENCY

This is more important than brute power in deciding radio range. Radio waves are a constantly switching mixture of electrical and magnetic energy, punched out into the atmosphere by the aerial. Once free, they travel at 186,000 miles per second – the speed of light.

The simple explanation is that in order

to liberate the signal-carrying waves, a voltage is driven up the aerial during the positive phase of the alternation. As it travels, it creates a field of electromagnetic energy outside the aerial. As this field collapses during the negative cycle, some of the energy remains out in the atmosphere and is punched away into space by the next positive pulse.

On Channel 16, which has a frequency of 156.8MHz this is happening over 156 million times a second. So, in order to cope, the antenna must be cut to a very precise length to suit the frequency. If it is wrong, the voltage which has run to the tip of the antenna (going positive) might crash into the next positive pulse as it runs back down again. The clash weakens and distorts the signal. If the antenna is too long, a void produces the same effect. This phenomenon largely, if simplistically, explains why you can receive on a poor aerial, but cannot transmit on just any bit of wire to hand.

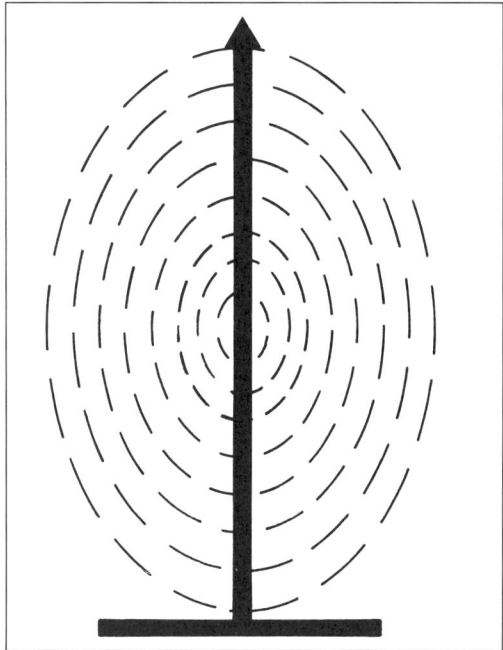

Electromagnetic field builds outside the antenna as the voltage goes positive.

## ANTENNA HEIGHT
Radio waves are said to travel in straight lines and that VHF communication is 'line of sight' because the earth is curved and you cannot look over its horizon to see another aerial. The obvious solution is to

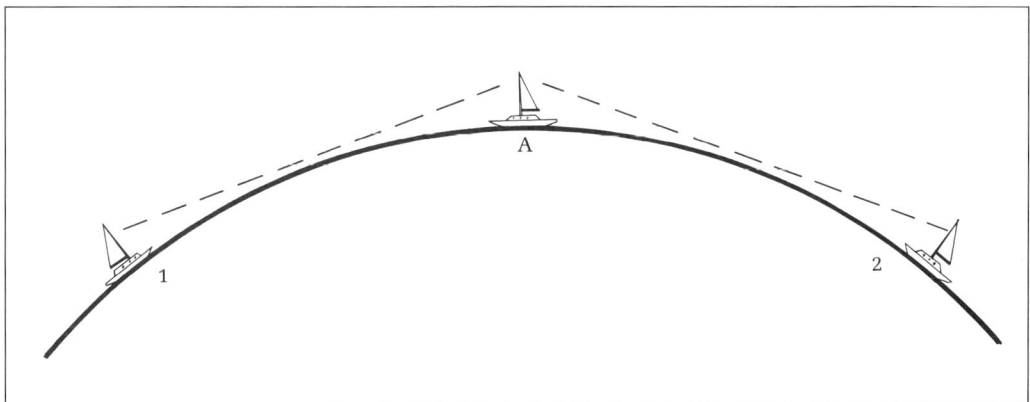

Boats 1 and 2 will both hear A, but not be able to hear each other.

put the antennae as high as possible so that the pair get a better look at each other, in the same way that you see a big ship's mast climb above the horizon before the hull appears. This limitation also explains why you can sometimes hear one station, but not the other to whom he is talking – he is probably over your horizon.

## PROPAGATION CONDITIONS

Sometimes called atmospheric conditions, propagation conditions also greatly affect radio range. Rain, fog and very low pressure can all reduce range, but a high pressure system causes 'ducting' and makes the radio waves follow the earth's curvature. This often enables a

signal to get over the horizon, and explains why British radio transmissions sometimes experience interference from the Continent and vice versa.

An experienced radio user will use such an uplift in range to reinforce other indications of a large area of high atmospheric pressure and probably calm conditions at sea for a few days.

# Electronic Barometers

These are now an important part of our on-board equipment. Not only do they give the present trend (and warn us that any sudden rise or fall means a blow coming), but they can also scroll back through the pressure readings of the past twenty-four hours.

Some models will display this on an LCD screen, but I personally find it more convenient to transcribe the back readings on to squared paper to show the twenty-four-hour picture. This has all the clarity and permanence of a traditional barograph at less than half the price and one-eighth the bulk, and it is immeasurably tougher.

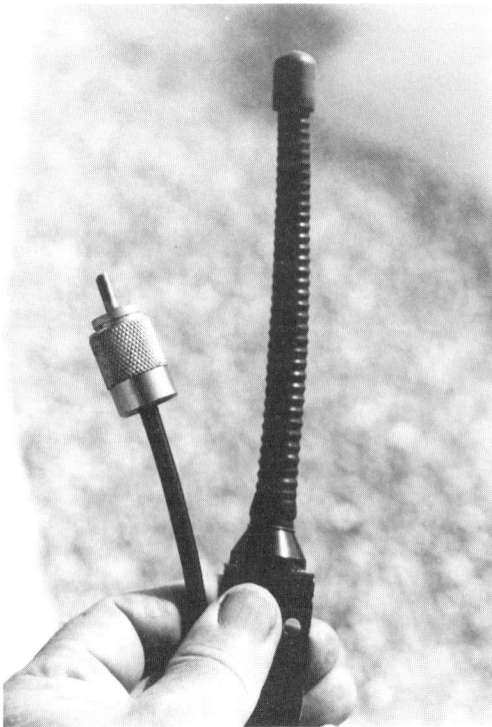

A stubby antenna will work much better at the masthead.

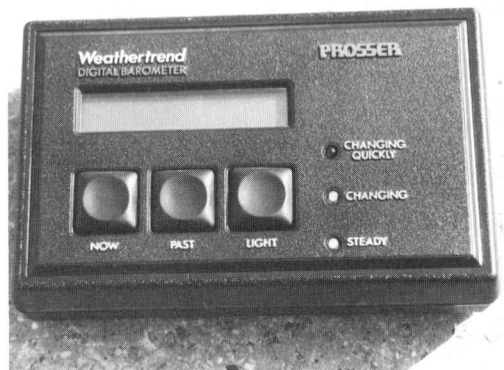

Modern electronic barometer.

## Navtex

Navtex also gives weather information, collated with navigation and distress messages, by various national authorities which broadcast on a common frequency. This is a low level way of getting an approximate forecast.

## Weatherfax

This is the high level way of getting a complete weather map (similar to television) printed at regular intervals throughout the day. The quality of the recorded image is variable, but it is unrivalled as a weather forecaster – as long as the operator learns to interpret the data. This scarcely comes under the title of a book aimed at beginners, but there are a number of books on the interpretation of pressure gradients and their effect on wind speed and direction. If your boat has weatherfax, they are essential reading and learning.

Weatherfax chart on small computer and paper.

## Speed and Distance Logs

These logs go together and are almost invariably in one instrument. Both pieces of information are generated by a paddle-wheel, itself activated by the boat's passage through the water. The spinning creates a voltage which varies with speed, and electronically converts it to knots and mileage. Leave aside the electronics and the principles are much the same as the speed and distance meters fixed to bicycles.

The major problem with such instruments is their lack of day-to-day precision. Only rarely will the factory settings be entirely accurate for a particular boat. The new owner must run over an exactly known distance several times in order to check if the meters are fast or slow. They can then be calibrated accordingly. An interesting, hi-tech way of doing this is to use the Decca or GPS. You could zero the tripmeter and note the time between waypoints, or use the in-built, resettable distance logs present in most electronic navigators.

Even when this has been done, however, you must make allowance for factors causing aberration. If, for example, the boat is stationary at anchor in a river, the speed log might show a couple of knots induced by the current. If you were proceeding up-river, the log might show 5 knots, when in fact the boat was only making 3 knots over the ground. Reciprocally, going downstream, the hull's effect on the paddle-wheel might be showing only 3 knots, but the extra push from the current would be giving 5 knots over the ground. There is also a difference between speed over the ground, which might be slanting, or crabbing, you sideways as well as forward and your velocity to target, in other words, how much closer you are getting to your actual destination with each hour that passes.

Once all the elements which you decide you need of the above equipment have been installed (it all depends on how far off-shore you plan to take your boat) you use their accumulated information to plan the shape of your passage to take best advantage of wind and tide, then use them again to monitor that the things forecast are actually happening, or whether you need to change your route plan. Even though I have a normal speed and distance log on the boat, knowing these factors in more accurate terms of speed and distance over the ground is so much more useful to the navigator than being shown them without effect of waves and current removed, that now I almost always take this information from the GPS unit on board.

# Radio Direction Finders

These are now rapidly being reduced to the role of 'back-up' navigation. Over the past two seasons, I have only used mine when I wanted to practise listening to the Morse signals to keep my hand in. I shall continue to keep it in the locker and to use it from time to time, but it has now largely been superseded by systems which are more accurate and more reliable. Its on-board presence is, however, essential as part of that belt and braces habit which all good seamanship demands.

**SUMMARY**

- The more your extras can back each other up the better.

- You cannot have too much electronic and screen information but do write it all down.

- Be aware of VHF radio licences and the penalties for not having them.

- Radio range and signal quality are 90 per cent governed by the aerial.

- Learn all the weather forecast rhymes which are old-fashioned but fun.

- Good extra instruments are chiefly installed to enable a dead reckoning plot.

- Get used to doing a dead reckoning in calm weather.

- The weather will always be of paramount seafaring importance, so a forecast method is essential.

- Weatherfax and navtex can now be received from many sources even without a dedicated system instrument.

- All the toys cost money – are yours safe from thieves?

# 9
# RADAR EXPLAINED

Radar decreases continually in bulk and relative price, but increases in quality and number of functions. It is a cruising tool which I hate being without, and I might well abandon a trip across the English Channel shipping lanes in poor visibility if I did not have it. Again, I can hear the Luddites screaming that they have always managed without it – just as we used to manage without electricity and heating. The reply is that if radar is available and you can afford to install it, the best possible advice must be to go out and buy it. Having radar on board makes you much safer and is lots of fun. There are many makes available for small boats, in a range of size, price and features; in recent years, the price has come down dramatically.

Basically, a radar set has two seafaring uses: collision avoidance; and navigation. However, in order to use the system to its best effect, you must understand a little of how it works, otherwise you will tune it badly and come to rely on a screen which is telling lies. You must also learn to use it properly, so that you avoid becoming what the marine trade calls 'a radar-induced collision'. Equally, it has to be realized that not all radar models are suitable for all boats.

In summary, a poor radar in bad hands is a very dangerous piece of equipment. Properly installed, correctly tuned and used by an operator who has taken the trouble to learn a few elementary skills, a radar set adds several extra utilities to being at sea. When you remember that professionals attend radar instruction courses covering several months before they are let loose at sea, it is surprising that many amateurs install a radar, forget about it when they are enjoying their sun-soaked gin and tonic, but panic when the weather palls and the radar screen does not give them the information they require, or more likely, they are not able to tune it, or interpret the received data.

To use radar well you must:

1. Understand the basic principles.
2. Spend time with the handbook for your model.
3. Get in plenty of hands-on, button-pushing time when the weather is very calm and you are not under pressure. Then, when the situation demands, you will do things correctly and, above all else, will be able to have implicit faith in what the screen is telling you is or is not there. This practice is essential in another respect. Not all boats show clearly on radar – wooden hulls and some in fibreglass materials make very poor reflectors, so they often do not return an echo until they are very close. Only by constant watching and button pushing will you acquire enough experience to guess when something might not show, and to interpret those echoes which appear weakly and then fade rapidly.

A radar has plenty of buttons to learn.

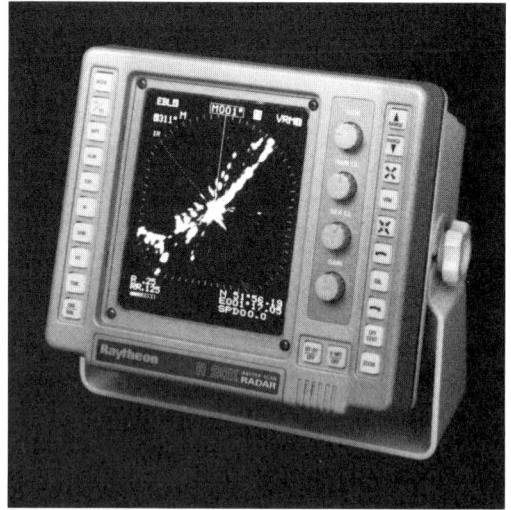

The raster-scan image remains on the screen.

## Basic Concepts

In many ways, radar resembles sonar because it works by shooting out pulses of radio energy, and not only times their return but also calculates the angle from which they were received. Because some surfaces and substances are better reflectors than others the on-screen images vary in size, shape and intensity. Interpreting these into an estimation of ship, trawler, yacht, land or buoy is just one of the tricks which comes with experience. In this, we are much aided because we no longer need to look down a tube at a blob which only appears when the rotating beam passes through. Modern radar screens have so-called daylight viewing, and are clear and constant.

Raster Scan screens have been the most significant advance of recent radar progress. By using pseudo-video techniques, they cause the image to remain on the screen even when it is not in a direct line with the rotating, transmitted radio beam. This makes life much easier for the operator, who can not only use the radar in daylight without putting both eyes to the rubber cover, but has much more time to mark a target's bearing and distance. This has the extra advantage in that he can keep one eye on the boat and the other on the radar plot.

Other benefits are the addition of bearing lines and distance rings produced electronically (rather than by use of plastic overlays), and the presentation of plenty of useful information in digital form. To assimilate this, let us look at various aspects of radar.

### Screen Size

This is measured diagonally, just like in television terminology. The bigger the screen, the easier it becomes on the eye and the more accurate will be your quick bearings taken from the 360-degree

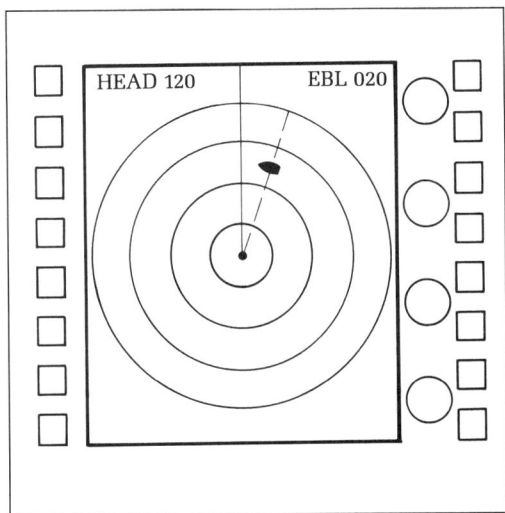

HEAD 120          EBL 020

Boat is on 120, target is 20 degrees off boat's head, look 140.

notation around the edge of the screen. This notation is used for the rapid location of a radar target by eye. If something shows as 20 degrees off the boat's head, you immediately know where to look. If you are in any doubt you do some rapid mental arithmetic. If the boat's head is on 090 degrees and the 'blip' is 20 degrees left, if you aim the hand-bearing compass down 070 degrees you should find him.

## Radar Display

The radar display is generally 'head up'. If you imagine yourself in a helicopter hovering above the boat, so that your view of the surroundings becomes two-dimensional, that is the picture which the display gives. An object at the top of the screen is directly ahead of you and an image showing at nine o'clock or 270 degrees is abeam of you to port. The angles in between are proportional. The helicopter analogy is not absolutely correct,

however, because radar only shows what the beam strikes first. If two ships are close in an overlapping line, only the first will show. Similarly, radar will not always separate a harbour wall from the land behind it.

## Output Power

This is measured in watts and is often quoted as the distance of the radar's maximum range. A yacht radar is not frighteningly power hungry, especially as a wise operator frequently runs it on stand-by.

To give an actual example which I have just tried, my own boat radar is a 40 watt 16-mile range model. At start-up it pulls 0.4 amps and continues at this level when it is on stand-by. When it is switched to 'transmit', the meter only moves to 2.9 amps of consumption.

## Range

Range is closely allied to power. To get more effective range, you need more output power. This is, however, something of a misnomer because radar signals, like radio waves, travel 'line of sight' and are therefore entirely dependent on antenna height. The higher you put the scanner, the further over the horizon the radar will be able to see. On the other hand, a 48-mile range radar with an aerial mounted on the coachroof of a motor cruiser, will only show images which appear above the short horizon such a location would permit. It would probably only receive its full range if mounted some 80m (250 feet) above a supertanker. Even then, though, one would seriously question whether the ability to spot objects at 48 miles is of any real use in normal day-to-day seagoing. The principal advantage of long distance,

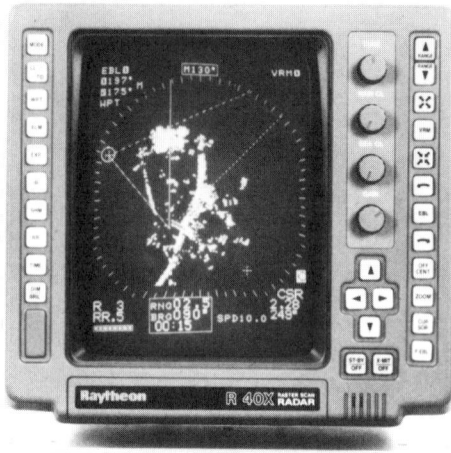

Display is head up; Guernsey on right and Herm to port.

high output power is that it gives a much clearer, sharper image at all the inter-mediate ranges. This is especially useful when navigating in cramped and busy places. The finer discrimination of big power separates targets which are close together. It explains why many boats which never leave the river Thames have massive 48-mile range radar installations, even though they mostly use them on the half-mile setting.

In my own case, the unit defaults to 2 miles at switch-on and I leave it there for

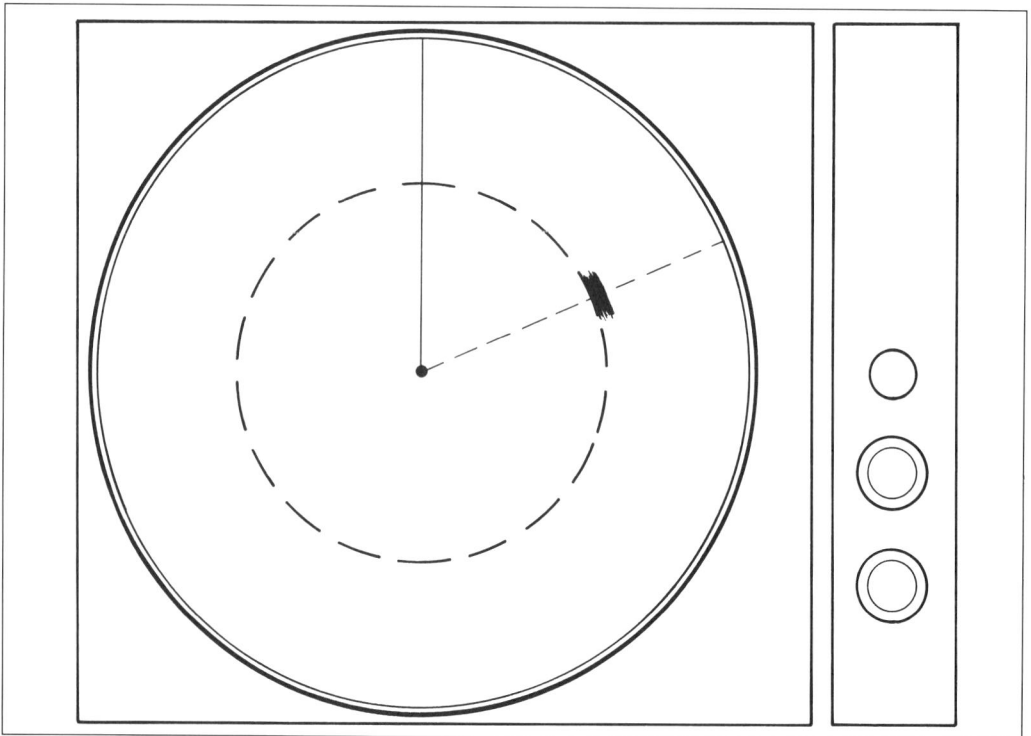

Radar display gives a helicopter view of 360-degree picture.

The open array is the one you see spinning.

coastal trips, with an occasional look at a mile or less if something interesting shows. Out at sea, I opt for a 4-mile norm, with an occasional look at 8 miles and cutting back to 2 miles or less if any target looks to be a potential hazard.

Power is always expensive both in terms of cash and space. The bigger radar will cost more than its smaller companion and the display unit will generally be both larger and heavier. The antenna will also be bigger and heavier. This limits many sailing yachts to an aerial which the mast and rigging will permit. Sail boats can

obviously put the scanner up quite high, which will give a better horizon. This is actually a mixed blessing, because once the boat begins to roll, a high radome might be arcing through 60 degrees or even more. It will not give a steady bearing in such conditions. As in all leisure boat installation, compromise is the answer.

## Antennas

The radar antenna is a combined transmitting and receiving device. It comes in two forms.

107

The radome is more usual for yachts.

the elements, but it does project a narrow signal beam.

## THE RADOME

The radome, or enclosed antenna, is more usually seen on pleasure craft. It comprises a printed circuit card array on two arms, spinning inside a lightweight ABS plastic dome. Its construction is usually in separate models to make replacement and service simpler.

The advantages of the enclosed radome are its light weight, simplicity of installation, streamlined shape and protection from the elements. On the down side, however, most standard radomes cannot take high power and they will not cope with big output power. The other minus is that they will not permit a narrow, precision target signal beam. Typically, my own boat's radar is offered with a choice of antennas. The more expensive model has an open array giving 24-mile range and a 2.5-degree beam. Our enclosed radome is reduced to 16-mile range and a 6-degree beam.

## THE OPEN ARRAY

The open array antenna is the one which you can actually see spinning as it transceives (it sends out signals and receives the returned echoes). In concept, it is a big boat device which, when carried to its logical conclusion, becomes the very large, slightly dished spinner used by warships. It is capable of handling vast output power and the shape concentrates it into a penetrative, narrow beam.

This description tells you much about the pros and cons of the spinning antenna, even on a yacht. It is the larger and heavier of the two, certainly more vulnerable to

## Beam Width

Beam width has an effect on the range of the signal but, above all, decides the relative size of the target shown on the screen.

Imagine a yacht 2 miles to the east, struck by the 1-degree beam of a warship radar. Its echo will appear on the screen as an exact pinpoint at 090 degrees, and this will actually change 1 degree at a time as the vessels pass and their relative positions alter. Such a narrow beam gives extremely precise bearings to the target even at long distances.

My own radar has a 6-degree output beam. As the array spins, the beam's

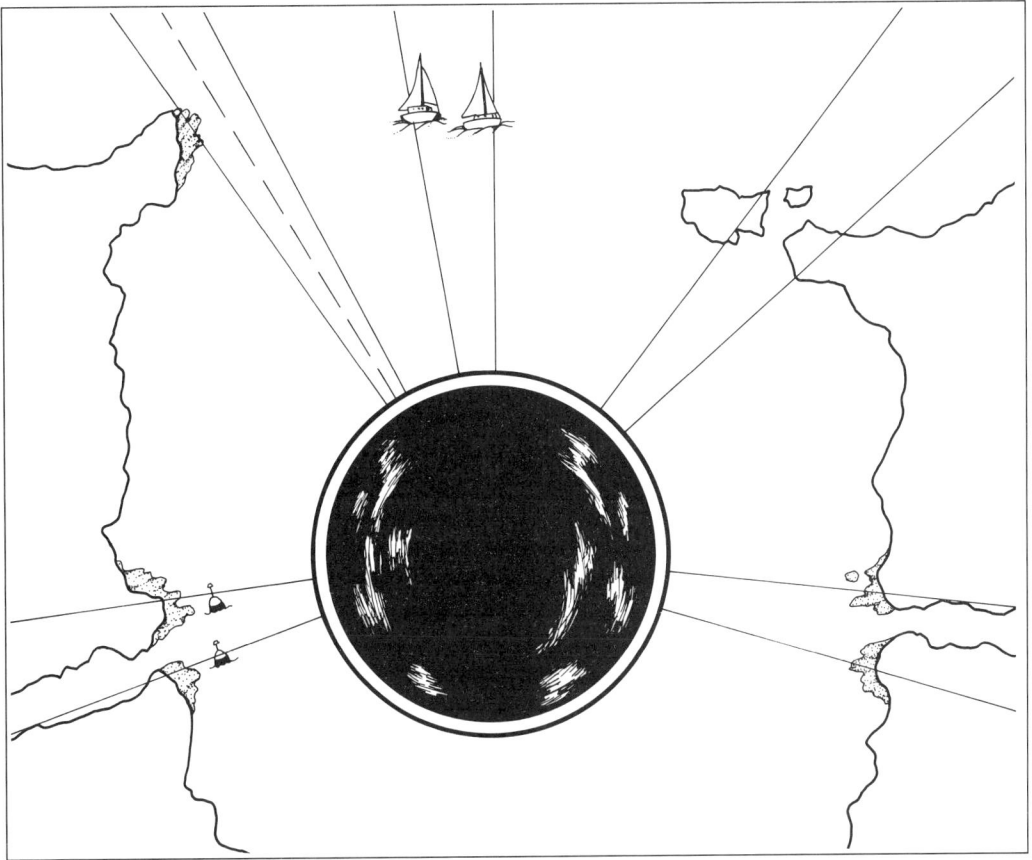

On screen, wide beam width merges boats and buoys.

leading edge would strike the yacht at the 087-degree medium point and would hold the image until 093 degrees is reached. This makes the target seem much larger than it really is. As the distance increases, the cone effect makes this even more exaggerated. The wide beam can also make two closely grouped targets appear as one, especially at a distance.

By choice, we should all opt for a narrow beam but, in practice, because radar is of most use to small craft at 2 miles or less, the cone effect is much reduced and targets easily separate. The bearing is also perfectly adequate for what most of us will wish to do. This is in turn decided by the functions incorporated into the particular radar installed. These vary considerably from one model to another but, fortunately for us, most manufacturers use a common language.

## Radar Speak

A knowledge of radar terminology and shorthand is prudent for two reasons:

1. You will not be baffled when a salesman launches into his selling spiel.

2. You will be able to understand the various controls (which are not always well explained in operating manuals), and will also be able to use all the functions properly, including the realization of their limitations.

## The Tune Control

This is an image refiner. It becomes more useful as the set gets older and the magnetron takes longer to warm up. The operator should then adjust the tune setting for the chosen range after ten minutes or so and again when the range is changed. Adjustment only takes a few seconds and is accomplished by picking a comparatively weak echo and setting the tune until it is at its clearest. Mostly you can use the land as a target.

## The Gain Control

The gain control determines the strength of the echoes on the screen. Its correct use is one of the most important skills to be learned. Initially it should be set for a lightly speckled screen background, then finely set to bring targets to their best image. With too little gain, weak targets will not show, or will disappear as their range increase. If gain is set too high, there will be insufficient discrimination between targets and background 'noise' – they will merge. Targets are then very difficult to pick out.

Lack of practice in balancing tune and gain is probably the most prevalent of all amateur operator faults – they generally set both too high. Only by constant practice will you get this right and only when you have it right will you be able to have total trust in your display. There is a tendency amongst some boat owners to think that they only need to buy a radar and switch it on to become collision-proof. It bears repeating that even though it is a very valuable piece of equipment, it needs more learning than any other. A badly tuned radar can, in fact, be quite a dangerous toy, for example, if it shows an empty screen when there is another vessel inside 1 mile. You only learn by constant button pushing – another reason to read the instructions for use, before you set off!

## The Sea Clutter Control

This control is designed to reduce the number of echoes returned by breaking wave tops. It is useful to clear the screen of 'junk' images out to about 3 miles. The control is most useful as you come down the range setting and nearby wavelets are showing at their strongest. At very short range, when the centre of the screen goes solid with echoes, the clearing effect of increasing the sea clutter is quite spectacular. It does, however, need using with a bit of care so that in clearing the screen of wave top returns, you do not also clear it of weak targets.

In effect, increasing the sea clutter reduces the radar's sensitivity. The operator needs to be assured that solid targets will still show. Again, the method is to select a weakish target – if you can find one – and then adjust the sea clutter until you can still see it, but have lost most of the big echoes near the screen centre. As a practical example with my own radar on its 0.25-mile minimum, by adjusting the control properly I expect to pick up lobster pot markers at about 300m (1,000ft) and even to see passing birds at that distance.

## The Rain Clutter Control

As its name implies this reduces the obliteration effects of rain and snow showers. Again, it must be used with enough discretion to minimize the interference without losing the targets. On the other hand, it is also fascinating to monitor the progress of scuds of rain as they pass within range.

Even in very fine weather, most radar sets seem to like just a small touch of both rain clutter and sea clutter.

## Interference Rejection

This reduces the on-screen effect of other radars within your catchment area. These effects show as curved lines radiating from the centre of the screen. I always set mine as a matter of course, after first switching on, warming up, and before inserting any navigational data.

## Target Expansion

This is also a useful on/off facility. Some echoes are too small for comfortable observation, for example, rowing boats, small yachts and navigation marks – so it helps to have their screen image enhanced. I also put the TE to 'on' and leave it there until I have a closely grouped set of targets and wish to separate them on the screen.

So far I have covered all the settings and functions which are regulated at tune-up. Once they are set, they can be forgotten until you make a radical change of range. The operating (or navigation), applications, however, are altered continuously to give the skipper the data he requires.

## Range Control

Range control alters the size of the area shown by the display. Generally it works in factors of two, doubling with each change from 0.125 miles to 0.25 miles, 0.50 miles, 1.0 miles, 2.0 miles, 4.0 miles and so on to maximum.

## Range Rings

These are concentric on the screen centre, or boat's position, and usually at 25 per cent of maximum set distance. At the 4 mile setting, a target touching the smallest ring will be 1 mile from the ship.

The purpose of range rings is to allow for a quick estimation of target distance, and they are an excellent guide when one is close enough to merit changing the range setting to get a better look. The range rings are fixed, but they can usually be switched off if the screen is getting a bit cluttered.

## Ship's Head Marker

The ship's head marker is that vertical screen line stretching from the centre – it extends from the antenna, through the bow and out to infinity. It tells the operator which side of the bow to look for a visual sighting. Mine is always left switched on unless I want a good look at something dead ahead which is being crossed by the SHM.

The point where the SHM line cuts the edge of the screen, is designated 0 degrees, and all target bearings are quoted in standard, clockwise 360-degree notation from this 'head-up' position. This is generally referred to as a 'relative bearing', so that it is not confused with a compass bearing. The exception to this

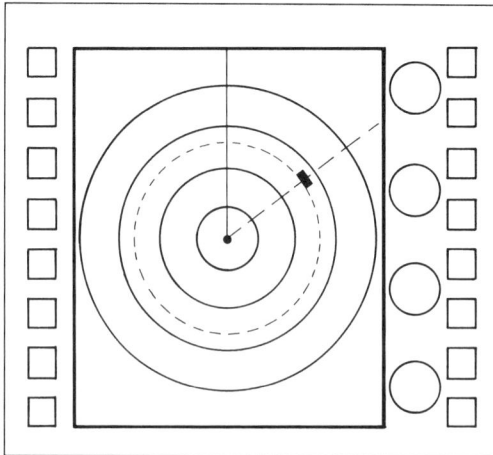

EBL cuts the centre of the target; VRM is on the leading edge.

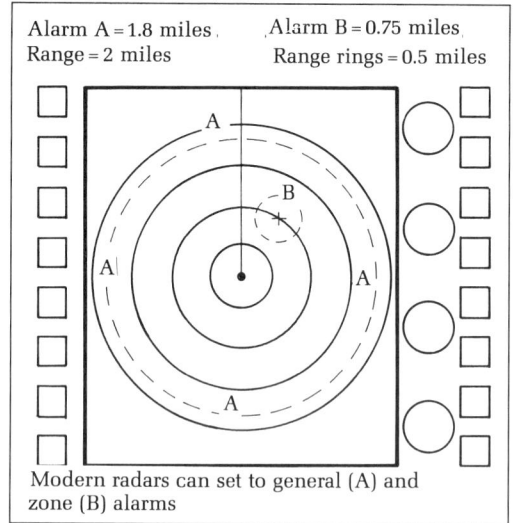

Alarm A = 1.8 miles,   Alarm B = 0.75 miles,
Range = 2 miles   Range rings = 0.5 miles

Modern radars can set to general (A) and zone (B) alarms

Alarm set for target coming into present zone, or our own boat going out of it.

rule are radars whose electronics can re-orientate the screen to show north up.

## Electronic Bearing Line (EBL)

The EBL is an on-screen pointer which can be swung in either direction from its pivot at the centre of the screen. When it is laid across a target, the relative bearing from the ship can be noted at the screen edge, or monitored from the digital read-out. Having two EBLs makes a lot of sense.

## Variable Range Marker (VRM)

The VRM is an electronic circle, expanded or contracted from the control panel, to cut the target and to give a digital display of its distance. The EBL and VRM are amongst the controls which you will use most often. The beam width phenomenon sometimes makes the bearing a bit imprecise, but radar measurement of distance is always incredibly accurate.

## Target Alarm

The target alarm is a function which allows the skipper to decide how wide a safety zone he would like around the ship and to set it on the screen. As soon as a target comes inside this *cordon de securité* an audible alarm sounds. The advantages of this are obvious. We should, however, also be aware that the dangers resulting from overdependence on electronic alarms, allied with a mistuned radar, are quite frightening – even with auto tune.

If a radar is integrated to GPS to show actual position, a circle can be set around the lat/long of a particular rock (or other point of interest) so that the alarm sounds when the boat comes within the pre-scribed distance – or passes out of it. Here, however, I am getting beyond my brief of writing a book for beginners and into the realms of boats with sufficient crew to dedicate one member to radar operation. The short-crewed, working skipper will probably also be radar operator, navigator, helmsman and lookout. He will develop many short-cut techniques and other habits to make life safe and simple. To appreciate these, I will now look at an actual radar-assisted passage.

## SUMMARY

- Radar is a super tool, but it is only as good as the operator.

- Most radars require constant tuning and adjustment.

- A magnetron will last for about 3,000 hours, so put the set on stand-by when it is not required.

- Keep clear of radar scanner radiations when the set is in transmit mode.

- Always have the means for converting relative bearings to compass direction ready to hand.

- Radar is very accurate on distance, but less so on bearings.

- Do not assume that a big ship's radar can see you.

- Do not assume that a big ship's radar is even manned.

- Learn the radar picture of your own harbour in good visibility. Draw it on paper and keep it ready for times when there is fog.

- Play radar but remember that your eyes are still better.

# 10
# RADAR AT SEA

Radar is possibly the most addictive of all the marine navigational aids. The ability to 'see' in the dark and to have eyes which pierce the fog, are two assets which almost put man on a plane with mythical gods. In days past, my own boat regularly crossed the Channel shipping lanes without the assistance of radar but, now that I have it, I feel a touch vulnerable if it is not available.

Unfortunately, radar is not only an expensive asset, but one of the most complex to learn to use well. It bears repeating that, if used badly it can be as much of a hazard as a life saver. Used well, it turns an amateur into a pro.

To remind you of the basic usage concepts already mentioned, radar is generally thought of in its two principal functions of: collision avoidance; and blind navigation. It can, however, do much more than these things, even in good visibility.

## Collision Avoidance

Most leisure boat owners fit radar because they plan to make an extended passage and wish to give themselves every possible chance not to become one of the disconcertingly high number of statistics of those who have either been run down by a big ship, or have been involved in a very near miss.

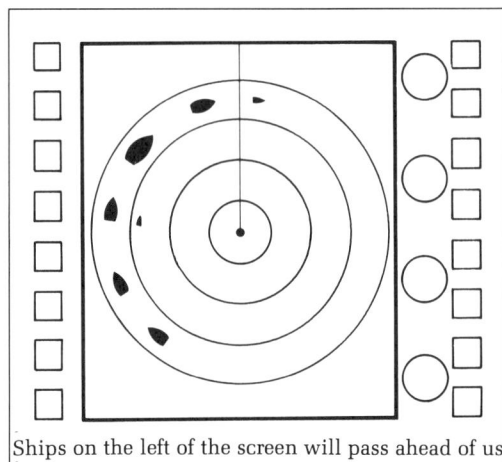

Ships on the left of the screen will pass ahead of us

Typical radar screen as we approach the channel shipping lanes.

They are, of course, very right to employ every possible means for such avoidance, but let us be quite straight from the outset and admit that radar alone does not make you immortal. Big ships generally seem to be closer than they really are. This fact causes many yachts to give way to them very early, and that is how it should be. One of the dangers of amateur radar, however, is that as the big ship's distance off can be tracked, the yacht skipper may delay whilst he is making observations. Then, the tanker disappears into the clutter at the centre of the screen, or creates such an enormous echo and false echo that it fills the screen and, suddenly,

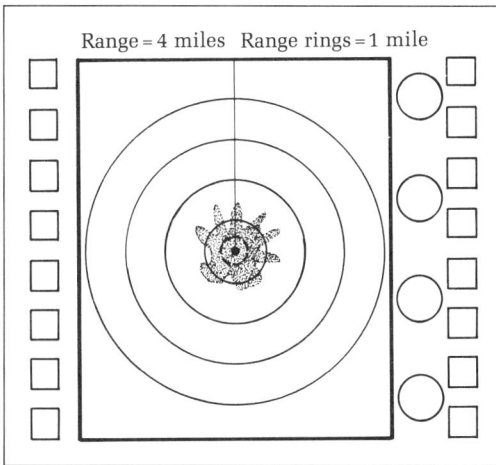

Range = 4 miles   Range rings = 1 mile

Danger point; a tanker has lost itself in a clutter around the centre of the screen.

the skipper does not know where the tanker is, except that he is close. Worse still, the skipper loses his mental picture of where the tanker is going and at what speed, because he has been screen-watching instead of eyeballing and comparing actual courses rather than electronic images.

That is the down side. The up side is that once you accept that radar does not relieve you of the responsibilities of good seamanship – including keeping a good human lookout and giving the other guy early warning of your intent – it is a superb collision avoidance instrument. Its efficiency in this mode is based on the things which radar does best:

1.  It calculates and shows the bearing between your own boat and any other vessel, or fixed mark.
2.  It is a very accurate range finder and measurer of distance.

The crew's function is to observe any changes in bearing and distance and to use this information to determine the other vessel's course and speed, and to calculate how close he will come to their own ship and what action is needed to keep both vessels safe.

## Simple Operation

Collision avoidance using radar is no different from any other form of seamanship. The basic rule remains: 'When seen from the position of your own boat, if the bearing of another vessel remains constant, but the distance between you constantly decreases, a collision will occur.'

In this sense, much good radar practice involves no more than 'keeping an eye on him'. In well over 50 per cent of cases, I simply note in which segment of the screen the other vessel is and his approximate distance as shown by the fixed range rings. After a couple of minutes I can mostly say 'No problem, he will pass three or four miles clear of us.' That, plus the fact that you can see the target and give an injection of skipper experience, will keep you out of trouble. The radar is here useful but subordinate.

If a target looks to be more interesting, I generally lay the EBL across him and put the VRM on as well. From these two, a bit of time will show whether he is coming down the EBL on a constant bearing, or whether he is clearing forward or astern. Yachts pose a couple of problems in setting these two electronic measuring devices.

First, my antennas wave about a bit up on the mast as the boat rolls. The target will appear to change places on the screen and sometimes will not even show up at all because the transmission is either

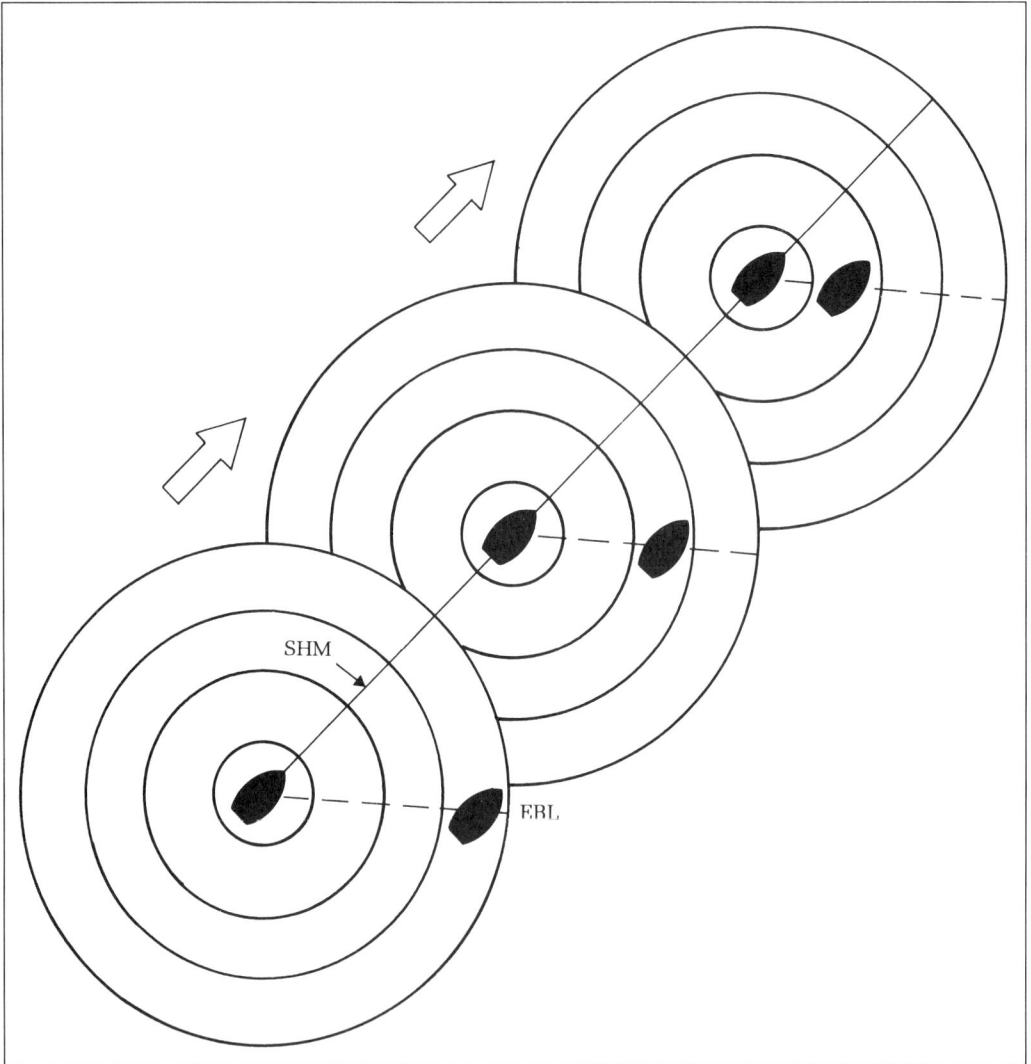

Bearing is constant; range is closing; collision is inevitable.

shooting up into the sky, or down into the waves. The lack of precision is also aggravated by the use of a target expander, which many screens need in order to display a very visible 'blob'. In favourable sea conditions, there are two things you must do:

1. Always let the EBL cut right through the very centre of the target.
2. Always put the VRM on its leading edge: the part closest to the centre of the screen. If the boat is moving wait until you get a good 'fix', before finalizing the placing of the electronic lines.

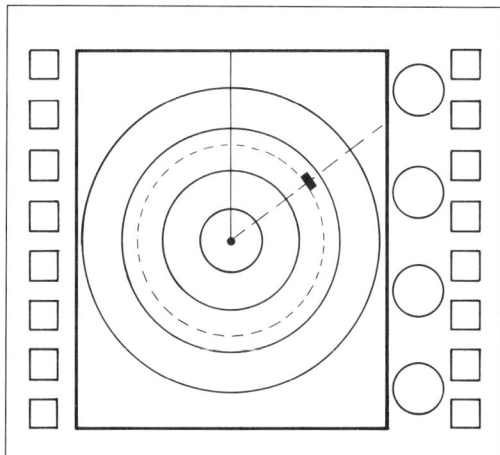

Target shows at side of screen, but will track around the VRM and pass ahead of us.

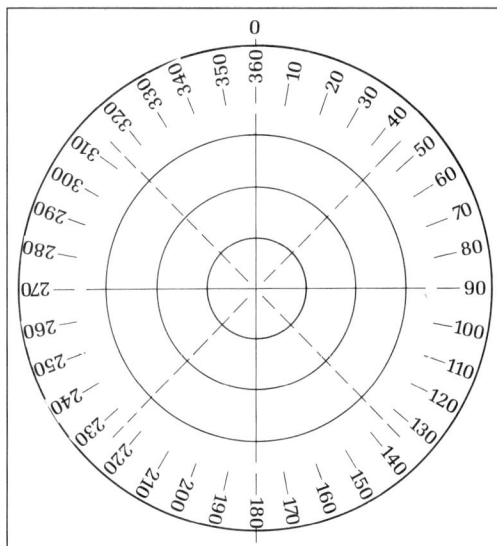

A spider's web, or simple plotting sheet.

Even in very good visibility, I keep a semi-serious radar watch, involving little more than a mental note of where the other traffic is. This pays dividends in big swell, when you momentarily cannot see another boat, even though you know that he was close but safe a couple of minutes ago. The same happens if the other vessel turns bows on and so presents a poor visual image, or when a fishing boat is weaving about to pick up his pots. If your eyes lose him for a moment, a quick glance at the screen will relocate him and tell you in which direction to look from your own boat to find him again. This is an example of radar being used in one of its best modes – as a subordinate assistant to the skipper.

## THE PLOTTED CPA

In a potential collision situation, the thing you need to know is how close the other ship will come to you and when. In radar parlance, this is his Closest Point of Approach (CPA). It can be determined with stopwatch, ruler and a plotting sheet, or a spider's web. The latter is an A4 sheet with concentric circles, just as you have on the screen, whose information can simply be transcribed to the paper. I always resort to recorded, serious plotting in poor visibility, or if there are many ships about at night and I want to identify the most hazardous, without zigzagging to confuse the others.

In its most elementary form, a paper plot begins at, say, 8 miles or 4 miles when the other vessel's range and bearing (taken from the EBL and VRM, or by estimation against the fixed rings and the screen edge bearing marks) are noted on the plotting sheet, whose rings are marked up to coincide with those on the screen. My own system is to try to get a fix down on paper every three minutes, especially at the closer distance. Obviously, six minutes, or one decimal tenth of an hour, makes for simpler mental arithmetic when calculating speed (for example, if he is doing

117

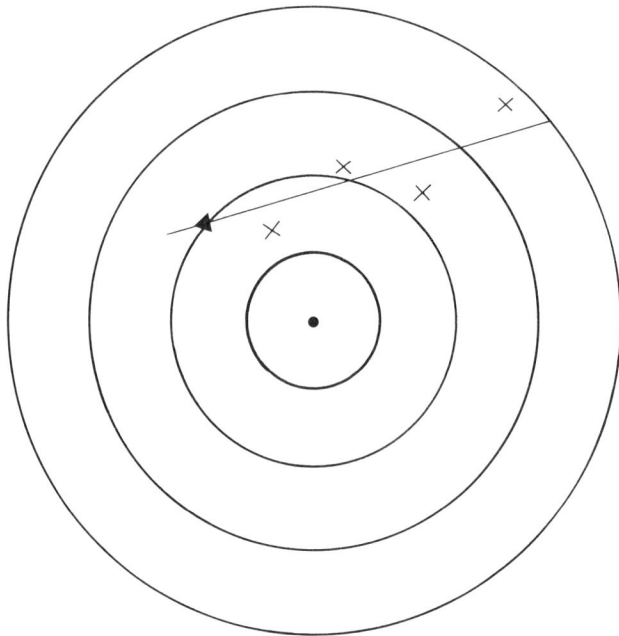

Take the median of the estimated plots.

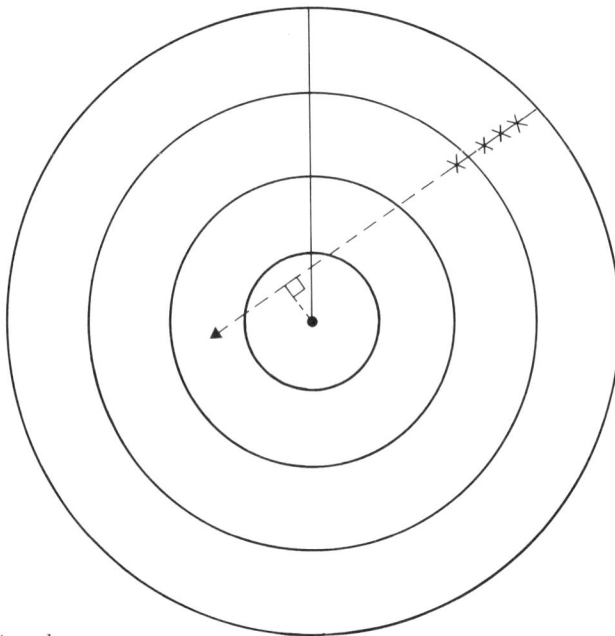

Project the plot line, drop a
perpendicular to your own boat, measure the CPA.

10 knots, he will cover 1 mile every 6 minutes. I prefer to exchange simplicity for the maximum amount of recorded data, and this is a preference shared by many other operators, who are not too proud to use an electronic calculator for the number crunching.

When you have a minimum of four plotted positions, you can join them up to show the other vessel's track relative to your own. It is more likely that the track will not show as a completely straight line on your sheet, so you lay a ruler along a median axis. When this line is projected beyond the centre of the spider's web (in other words, your own position) it will show how the other vessel will pass you – ahead, astern, or down one side.

To calculate how close he will be, drop a perpendicular (right angle) from the track line to the centre spot and measure this off against your range scale. To this, you can add the time when your target will actually reach this CPA by extending your stopwatch timings. An easy way to do this is to put the dividers across one of your 3- or 6-minute fixed intervals and walk them on down the line to see how many more intervals will occur before he passes you. This plot does, naturally, assume that neither your boat, nor the target vessel, will change direction or speed. You can also plot directly on to the radar screen using a non-permanent 'chinagraph' pen, but be warned that this can become a little messy especially if you run the radar long enough for it to get very warm.

Such a simple plot will mostly give you all the information you require to avoid a collision. I find that I can do this when I am single-handed with the boat on the autopilot. It is, however, much easier if you are double crewed, so that one either calls out the bearing and distance for his companion to plot, or you can dedicate one watchkeeper to navigation and radar plotting.

In nine cases out of ten, if a collision seems likely, it will be the yacht who will take the avoiding action, even if it has the right of way. It is a mistake to assume that a big ship will have a separate radar operator, or that you will show up on his screen. I know several trawler skippers who will refuse to change course for yachts with right of way unless collision is unavoidable at the last moment. There are also many merchant fleet operators who cannot interpret small blips on the screen. A personal acquaintance, who was Mate and Senior Watchkeeper on a 'flag of convenience freighter at the raw age of nineteen, confesses that it was only the biggest echoes which concerned him.

That is another excellence of radar. It is a super piece of self-preservation gear, enabling you to take your evasive action early, positively and correctly. The need to do this might even persuade you to run a full speed and distance plot – worth doing in good visibility, not only for fun, but also to make you better understand your own strengths and weaknesses, in addition to those pluses and minuses of the electronic aids.

THE FULL RADAR PLOT
Maintaining a full plot is just a little more complex, but it is very interesting to do and has, on occasions, been a real life-saver to a boat with limited ability to manoeuvre.

When you make your first 3- or 6-minute plot, imagine that your target has simultaneously dropped a marker which will remain stationary. As seen by your radar, it would appear to move vertically down the screen, parallel to the ship's head marker, as your boat moves along the SHM track. You and the imaginary marker will be like two buckets passing each other in a well. The speed of the marker's descent will, of course, exactly match your own speed over the ground and can be divided up into 3- or 6-minute segments just like the target plot.

Take a minimum of five plot spots if possible and also note down five imaginary marker spots as illustrated on page 118. Then a line drawn from the marker's fifth plot spot to the fifth position plotted from the target will give the other vessel's true course and speed, which is data to add to the CPA, which you will obtain from your projected line. This is more clearly illustrated with a couple of examples.

## Example One

In the first, I have a target which I first pick up at about 5 miles and set myself up to plot him as soon as he comes within my 4-mile range. As soon as he shows on the 330-degree line, I draw in the crosses for his imaginary marker. They are placed 0.5 miles apart and noted as being every 3 minutes. This echoes my own boat speed of 10 knots (1 mile every 6 minutes, or 0.5 miles every 3 minutes) and brings the sixth cross abeam to port.

Over the next 18 minutes I plot the target's position on my spider's web, exactly as I observe it on the screen. I see that he is coming down a line 330 degrees to myself, and the VRM shows him to be moving 0.25 miles every 3 minutes. In the 18- minute plot duration, he has come 1.5 miles closer. If I draw a line from marker six to plot six and project it, I will see that it slants ahead of us and across.

The conclusion is that I am overtaking a slower boat, who is also converging with me, but it still remains my duty to give way to him, because I am the overtaking vessel.

## Example Two

In this second example, my own yacht is doing 5 knots, but for continuity's sake, let us again imagine that I pick up the target again on about 330 degrees and start plotting him as soon as he is within 4 miles.

The crosses to denote his marker are brought down the plot sheet to show my speed, which is 0.25 miles every 3 minutes. The other vessel is plotted and seems to be getting closer at the rate of 0.5 miles every 3 minutes, or 1 mile every 6 minutes. (Note that this is his closing speed, which can be very different from his speed through the water.)

Again I continue the plot for 18 minutes and project the line which joins marker six and plot six. It shows that I have a vessel slanting towards me and that he will cross close ahead. By putting the dividers across one of my separate 3-minute plots and comparing this with the distance from marker 6 to plot 6, I can compute his speed.

Technically, he is showing my 'green' and should be able to see plenty of 'red' from us. It is my duty to stand on, to hold my course and speed, so he should give way if there is any risk of collision. As it is, I shall keep a good lookout from the window, as well as watching the radar. Then I shall be able to take early evasive action if the approacher is very large, or more interested in his gin and tonic than in good seamanship.

As I am making up my mind, I shall also be reciting one of the many versions of an old sea rhyme.

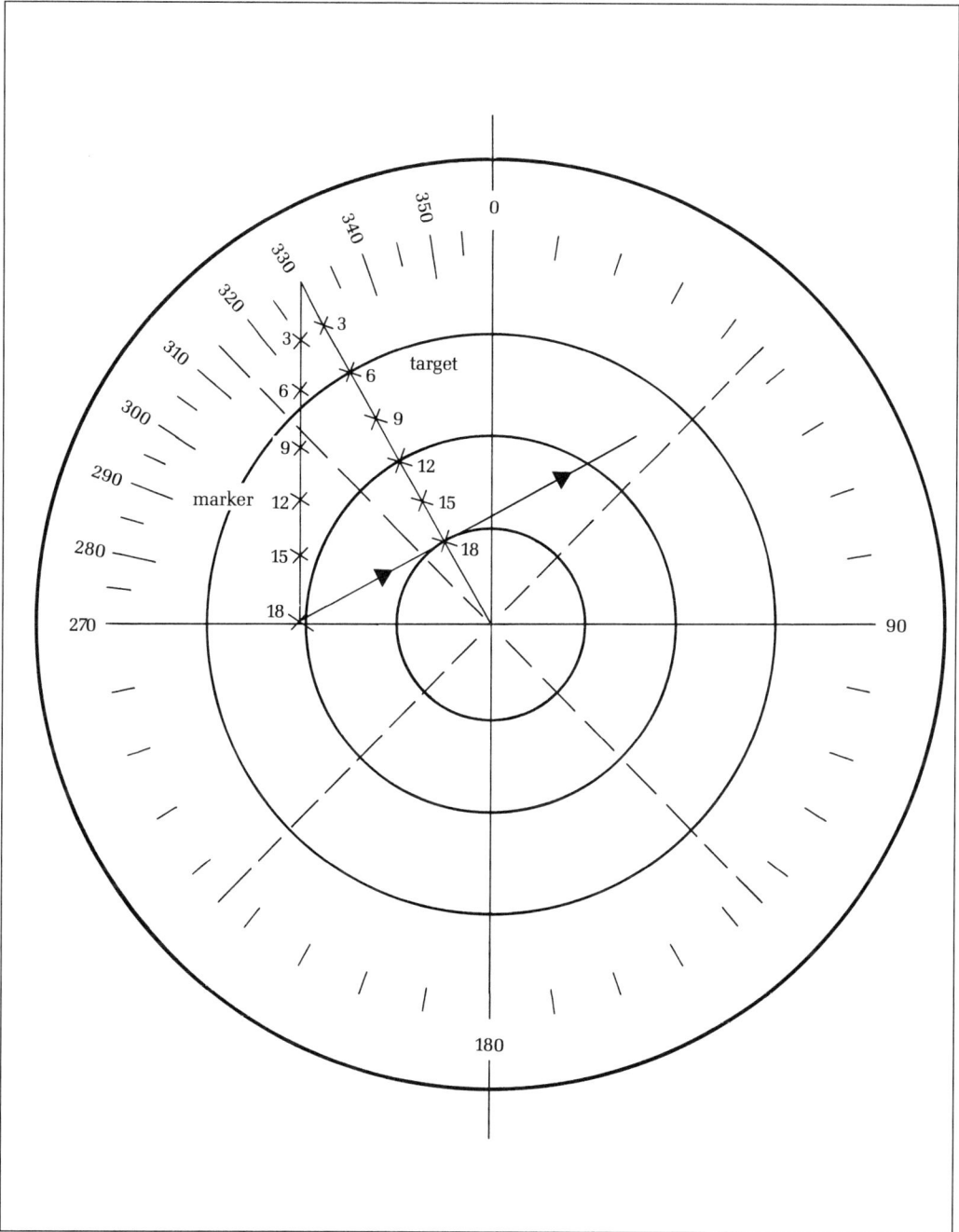

Example 1.

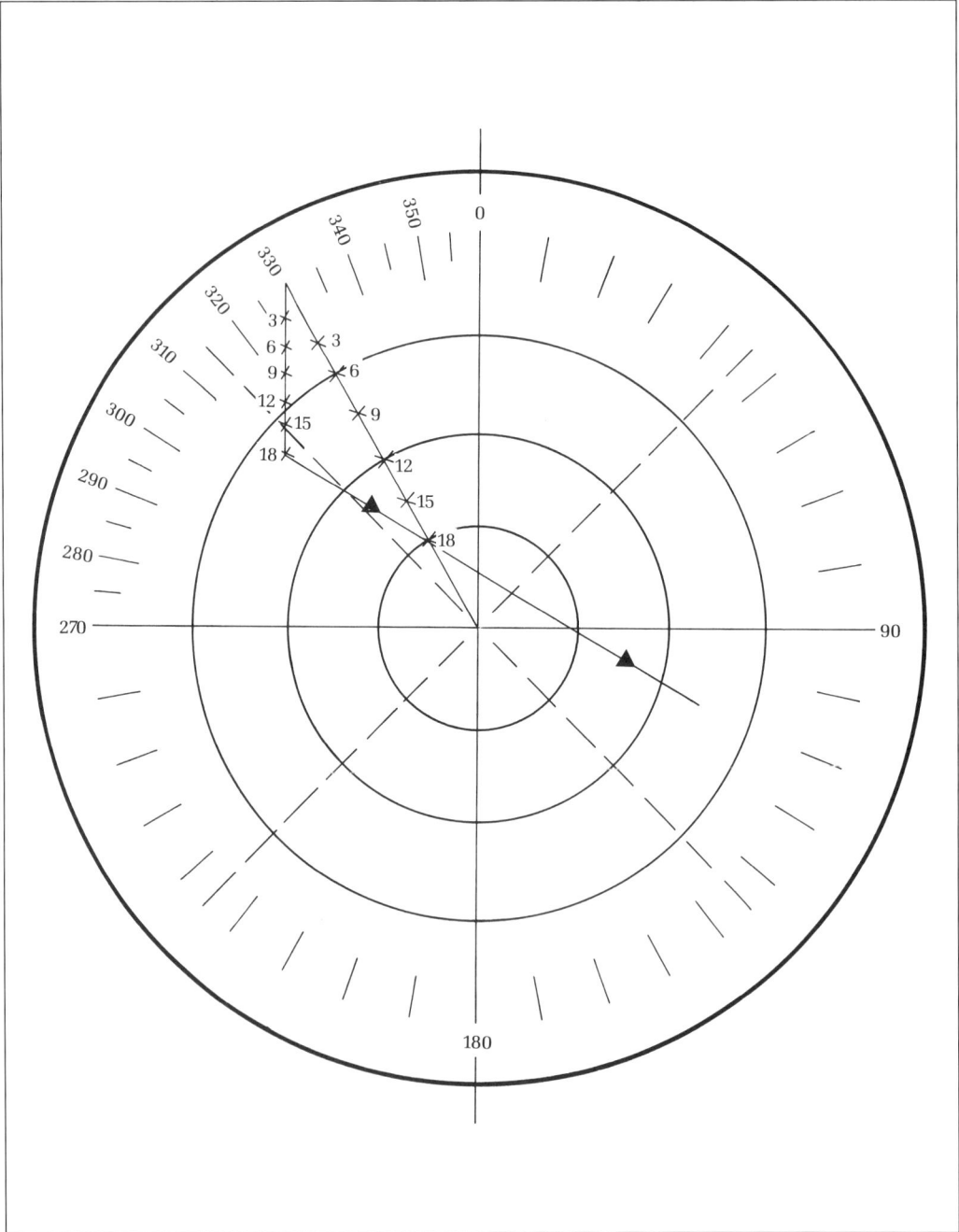

Example 2.

*This is the tale of Captain Grey,*
*Who knew that he had the true right of*
*way,*
*He refused to budge and just travelled*
*along,*
*But he's just as dead as if he'd been wrong.*

In crowded areas like The Solent, many potential collision cases are quite clear cut. The two protagonists are often sailing yachts or power cruisers, which are at all times subject to the anti-collision regulations, and it makes for sense and safety to obey them. This means doing the correct thing and doing it early enough to give the other boat a clear indication of your intentions. You must, however, also remember that in addition to international rules many ports have very precise local regulations about where yachts should travel, so that they do not expect a large and slow turning merchant ship to alter course – either quickly or dangerously.

In both cases, the advantage of running the radar, even when the visibility is good, is that you have a more accurate picture of the path between vessels than you ever get with your eyes. So, you use both eyes and radar. This combination really comes into its own out in the shipping lanes of the open sea. It is also out there that most skippers meet problems and have to make decisions which might not be covered by the anti-collision regulations. Oddly enough, as many serious collisions occur well out to sea – where there ought to be plenty of room – as in crowded port approaches.

## Radar and Big-Ship Lanes

My own experience of radar is probably typical of most leisure users. I do not like to be without it, but I also came to learn its limitations and to see it in perspective as just one more aid to sea safety (which is useless on its own) rather than as a total navigational and anti-collision panacea.

The major problem is always big ships which come too close – or which we allow to come too close. Up on a tanker's bridge, steady even in a moderate sea, the operator of a powerful, narrow beam radar can often calculate that he will miss a yacht by 0.25 miles – or sometimes even less. Aboard the yacht, and even aboard a yacht with radar, life is never that simple and never that precise. We cannot work to such close tolerances, so it is always best to break away on to a safe course if a big ship looks as if it will come close. This applies especially to those which the radar shows to be overtaking us, or converging astern at a shallow angle.

Perhaps the best anti-collision advantage of radar is that it shows vessels coming up behind – a direction where many yachtsmen never look. The danger is that because you can see him on the screen, you let him come too close and then you both alter helm in the same direction. Having had one or two pulse-raising incidents, I now break away early for all big ships whether I have right of way or not. I have given up calling radar targets on VHF radio. There are either language problems or no answer, or – as happened with a Danish car transporter about to come into my wheel-house in the fog off Ushant – the lady watchkeeper announced 'I have no intention to alter course.'

## Radar in Blind Navigation

I have recently made a number of passages which I should have aborted, or would have been forced to heave to had I not had radar aboard. I must, however, also say exactly the same about any of the other instruments in the navigational orchestra – GPS, Decca, echo-sounder, wind indicator (flag), charts and an observant crew.

The advantages of radar are that it extends your navigation possibilities, makes life a bit easier by confirming data from other sources and, occasionally, makes life safer if you have to run for port because of worsening weather or visibility.

Right from the start, it pays to look at the radar screen very often as you enter your home port in excellent visibility. By this means you retain a mental picture of what shapes line up with each other on the screen at different points of approach. Some wise skippers even draw rough plans of what they see and keep them in the cockpit notebook. Then, when the visibility is bad, they are absolutely certain of the correct interpretation of the overlapping images before them. The idea of having a notepad to confirm a radar screen, which is in turn confirming a GPS plot, itself backed up by depth sounding, is neither a crazy idea nor overkill. Good navigators always want the maximum amount of information. Quadruple confirmation makes sense and radar is an increasingly important element of it.

### SUMMARY

- Practise using the functions and pushing the buttons on good visibility days.

- Draw and photocopy a supply of plotting sheets.

- Learn the principal functions so that you can perform them even in the dark.

- Get to know the less usual functions, so that if you switch them on by accident you know how to switch them off again.

- Your memory is not good enough, write down all bearings.

- With practice, targets can be interpreted – trawler, yacht and so on – but only with practice.

- Learn how to do a very quick plot on a jotter pad.

- Work in units of six minutes which equal 0.10 of an hour.

- Cluttered screens are very dangerous; switch off all superfluous information.

- Having radar makes you safe, but it does not make you immortal.

# 11

# THE FUTURE OF ELECTRONIC NAVIGATION

So, where do we go from here? Forecasting is ever a dangerous and fascinating game, but it is easier here because the sea never changes – she will always remain a charming playmate and a vicious mistress when things go wrong. In this, the sea cultivates the need for high standards of safety, navigational excellence and the security of an alternative backup. This means that seafarers adopt new ideas and hardware only after long periods of reliable testing – a fact which explains why some highly innovative electronics seem to take a long time to become popular. There are, however, a number of very certain facts.

Most European countries have pledged their intention of retaining and improving their Decca navigation systems well beyond the year 2000. This system has proved to be very reliable in the past and will still serve most of our needs. Inevitably, the principal blue-waters position-fixing system will be satellite based. The US GPS system is already working well and it appears likely that others will join it. With equal certainty, Differential GPS, which adds a shore station to the satellites, will also be more widely available and lower in price. These shore transmitters will be based at major harbours and will give phenomenal accuracy for guiding us in during fog. It is already possible to integrate and interface most navigation, radar and autopilot systems to work with each other, as described earlier in the text. Whereas the ability to have the same information display in various places is obviously welcome, the use of such functions as a GPS navigator totally guiding the ship via the autopilot obviously poses certain dangers to small craft. For this reason, its progress into much use will inevitably be slow.

On the boat's bridge, we shall all become more used to reading screens instead of paper pages and will certainly need to become more computer literate. The black box will continue to get smaller and each model will offer more information. Conceivably, there will be a satellite navigator combined with a weatherfax decoder, including a software program which interprets the charted data into a verbal and graphics visual forecast for your current position. Graphics still have a long way to go, in spite of the

Plotter graphics will soon catch up.

enormous progress made to improve the quality of trackers, plotters and electronic charts. At the moment, the screens are simply not sharp enough, nor large enough to compete with the fine detail and panoramic information displayed on a paper chart. They will, however, eventually catch up.

Will this happen eventually or soon, for who can believe the speed of electronic development? Just be grateful for the way in which progress has made our seagoing lives easier, safer and a lot more fun. Those of us who admire the simplicity and easy operation of the best of marine electronic systems, certainly live in very exciting times.

# Index

# INDEX